Trench Rescue Awareness, Operations, and Technician Levels

Trench Rescue

ISBN: 1890306428

Library of Congress Control Number: 2002115532

www.trenchrescuebook.com

Photo Credits

Spec. Rescue International, Virginia Beach, VA.
Airshore International, Steve Cudmore
Paratech, Nigel Letherby Technical Coordinator
Chuck Wehrli, Captain Naperville Illinois Fire Department
Bob Schip, Flemington-Ruritan First Aid and Rescue Squad

Illustrations Credit

Steve Campbell

About the Author

C. V. "Buddy" Martinette, Jr., is the Chief of the Lynchburg Fire and EMS Department in Lynchburg, Virginia, USA. Prior to this position he spent twenty-five years with the Virginia Beach Fire Department, leaving as a Battalion Chief and Chief Fire Marshall. Chief Martinette has a Bachelor of Science in Fire Administration from Hampton University and a Master in Public Administration from Troy State University. In addition, he is a graduate of the National Fire Academy Executive Fire Officer Program. Chief Martinette has received the designation of Chief Fire Officer by the Commission on Chief Fire Officer Designation.

Chief Martinette is an Instructor IV with the State of Virginia Department of Fire Programs, Incident Support Team Operations Officer, and Task Force Leader for Virginia Task Force II of the Federal Emergency Management Agency's Urban Search and Rescue Program (US&R). His Urban Search and Rescue experience includes the Colonial Heights Wal-Mart Collapse, deployments for hurricanes Floyd and Fran, the Murrah Federal Building Bombing, and the more recent Pentagon collapse.

Chief Martinette is an US&R "Rescue Specialist Instructor" and lectures nationwide on specialized rescue operations to public safety, military, industrial, and law enforcement organizations. Chief Martinette has been active in course development for all areas of technical rescue and is best known for his course development and delivery in the areas of Confined Space Rescue, Trench Rescue, and Structural Collapse Operations.

Acknowledgments

To Sarah, Brittany, and Ryan who have spent many hours without their husband and father so he could follow his passion of learning and teaching specialized rescue operations.

CONTENTS

TRENCH RESCUE

AWARENESS, OPERATIONS, AND TECHNICIAN LEVELS

BEFORE YOU GET STARTED

This manual is not stories about my life. It is a lot of words that describe how and what I have taught students over the years about trench rescues. The problem is that I cannot lay claim to all these words. Many of them, as a matter of fact, most of them, were taught to me at some point in time. The end result of mentoring and learning through teaching with some of the world's most capable rescuers and accomplished instructors is what is contained in the first edition of this manual.

If you are looking for a manual that will give you ideas and tools, then keep reading. You are going to get a bunch of ideas and tools. It is my desire that when you complete this manual and receive additional practical training, I will get that call or E-mail about a successful operation from you.

COURSE DESCRIPTION

This Student Manual is intended to be an instructional aid for the awareness, operations, and technician level trench rescue course. The purpose of this course is to familiarize fire and rescue personnel with procedures to conduct rescue operations that will mitigate most types of trench collapse incidents.

The Trench Rescue Awareness Level is the first of a three-module series and was designed to give the student an overall awareness of trench rescues, and to support a cognitive background on which to build to the Operations Level.

The Operations and Technician modules will familiarize the student with the equipment and techniques necessary to effectively render certain collapse situations safe for the rescue of entrapped victims.

See Unit 1 page 10 for specific information concerning appropriate units for Awareness, Operations, and Technician Levels.

DISCLAIMER: A trench rescue can be a dangerous and potentially life threatening activity. The information contained herein is based on currently accepted trench rescue practices and theories. **This material should not be interpreted**

as a substitute for training by competent and experienced instructors. The manual is intended to complement additional hours of hands-on practice in actual trenches.

OBJECTIVE

The overall objective of this trench rescue manual is to provide you with as much trench rescue information as possible, once again stressing the "tools for your tool box approach." With this in mind it can be assumed that more tools will equate to an enhanced success rate when you attempt to make a rescue.

What this means to you is that all of the techniques, procedures, and suggestions mentioned in this manual will work, but not in every situation. In fact, what might be an excellent technique in one situation may make you look like a complete buffoon in another case.

My task then is to expose you to as many ideas, procedures, and techniques as possible so that you stand a good chance of success when situations look different from what you had anticipated.

As we journey down the road of techniques, keep in mind there are many ways to do the same thing correctly. The individual techniques and procedures discussed here are not the only ones that can be used. Additionally, we could argue until the cows come home that this one thing should take place before another thing. (Typical behavior for the tech rescue nut who is always looking for ways to make the job easier and for things to argue about.)

The bottom line is this: don't get bogged down with technical procedures, or with someone who tells you there is only one way to do anything. Just because this manual says to follow these steps, don't for a minute think that someone else doesn't have a set of some different steps that would be equally successful. These are just the ones that I have found to be successful, and that I know for a fact will, if you follow them, lead to success.

SPECIAL NOTE TO INSTRUCTORS

If you are an instructor using this manual as a teaching tool, the following is suggested to ensure a safe and effective learning experience for the student.

PERSONAL PROTECTIVE EQUIPMENT NEEDED

Students operating shoring equipment for familiarization purposes should be required to wear personal protective gear, including a helmet with eye protection (no burk eye shields), hard soled/steel toed boots, leather work gloves, and a jumpsuit, or appropriate pants and shirt. If the equipment lecture portion of the class will not involve student participation, no personal protective clothing or equipment should be required.

FACILITIES NEEDED FOR LECTURE

- A classroom location with chalk board or white board

- An overhead projector (as necessary)

- A laser pointer

- Video projector, screen and remote control device (as necessary)

- 1/2" VHS VCR and monitor (as necessary)

FACILITIES NEEDED FOR PRACTICAL

- See Appendix Two

PowerPoint slides are available for the Awareness, Operations, and Technician Levels at www.trenchrescuebook.com

UNIT ONE

CONSIDERATIONS FOR SPECIALIZED OPERATIONS

TERMINAL OBJECTIVE

To determine the considerations that make specialized rescue operations different from traditional fire and rescue work.

ENABLING OBJECTIVES

The student will be able to:

- Define the term technical rescue as it applies to the "Big Three"

- Discuss the rescue training cycle as it pertains to specialized operations

- Identify the four service levels associated with all technical rescue operations

Trench rescue is one of many disciplines associated with the term "technical rescue," which is a generic term for special rescue operations requiring the **"Big Three."** The **"Big Three"** includes **special people, special equipment, and special training.** Failure to integrate these elements into an active program for trench rescue efforts will result in a weak and potentially flawed system.

The types of personnel who comprise any non-traditional rescue team, (trench, confined space, structural collapse, rope rescue, etc.) are much different than traditional fire and rescue service providers. They truly are **special people**. Special for their ability to operate in a highly dangerous, unforgiving, and unpredictable environment with limited resources. Additionally, we expect them to think clearly in situations that, from a patient survival standpoint, may ultimately be hopeless, while requiring them to maintain a good attitude, be subjected to peer criticism, and endure intense training in order to maintain proficiency in situations that do not occur very often. As we discuss special people and teams further in the text, you will see that in every case, the foundation of your success in specialized rescue will be rooted in the people that make up the team.

If the most important aspect of your team is its membership, by far, the second most important part of the "Big Three" is **special equipment**. It is vitally important to the rescue effort that the rescuer is provided with the special equipment required to do the job safely and effectively—specialized, highly technical equipment that is difficult to maintain and expensive to operate—the type of equipment that will ultimately require the rescuer to analyze a situation first and react second. Like the difference between an atmospheric monitor and a fire nozzle. Get the picture?

Now, with that said, if you are reading this manual because you are interested in getting involved in trench rescue, but you are not willing to support that effort with the proper equipment, stop now. All of those great people on your team are destined to fail. Think about it. You wouldn't expect a doctor to do open heart surgery with a hammer, would you? Make sure your team members have a reasonable chance at success. If you are going into the specialized rescue business, get the specialized rescue equipment required to be safe and effective.

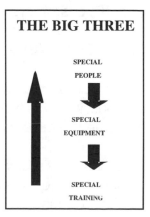

THE BIG THREE

SPECIAL
PEOPLE

SPECIAL
EQUIPMENT

SPECIAL
TRAINING

The third element in the "Big Three" is **special training**. Specialized training is necessary because all of those special people and their special equipment have got to function together. We are not talking about a "run of the mill" training effort, but, instead, a training program that is solid, realistic, and practical to provide the skills knowledge and abilities required. The bottom line is that the special equipment involved will seem foreign when it is needed if you haven't taken the time to train. If you are doubting me at this point, run out and ask someone to set up the high-pressure air bags or extrication equipment for time, just like in an emergency. Generally, what you will find is that people think they know their equipment, but when the pressure is on they fumble around and look at it like it came from outer space. To prevent this phenomenon, practice often and hard. Challenge your people so that when the chips are down and all elements are working against you, lack of training will not cause you to fail.

THE SPECIALIZED TRAINING CYCLE

The development of the "Big Three," as it applies to teams, is a constant circle of evaluation. The people you choose as rescuers will be continually recruited and their skills developed. The equipment you purchase will need to be continuously evaluated and updated, and thus frequent training provided to support the system. Make certain that you invest equally in all aspects of the specialized rescue cycle if you want to be successful.

For the application of this manual, our efforts will be to combine the "big three" to safely and effectively rescue individuals that may be trapped by collapses and/or any other form of medical/trauma emergency occurring in a trench or excavation environment. In doing so we are setting forth on a journey to learn trench rescue and are committing to providing the most effective customer service to both our internal (fire and EMS personnel) and external ("poor stiff in the trench") customers.

SERVICE LEVELS

The delivery of technical rescue services, and in this case trench collapse operations, involves a "layered or integrated system" which addresses the needs of all customers. These levels are broken down into four specific areas which include:

- **Awareness Level:** This level provides information for the first responder to identify the hazards associated with collapse and its associated dangers. Awareness training provides a decision-making matrix that allows first response personnel to begin the process of incident stabilization. Awareness level personnel are not intended to be actively involved in the rescue operation. Instead we educate them so they become part of the solution, and not a part of the problem.

- **Operations Level:** Operations represents the first level at which personnel learn the necessary techniques to render certain types of collapse environments safe for subsequent rescue operations. In some systems, these personnel may also be responsible for the initial, or long-term rescue operations. These individuals are called **"support personnel."** Placement of sheeting and shoring systems in all trenches that are not more than 8 feet deep and do not intersect is within the operational scope of these personnel. Personnel at this level may also function as sector officers during the development of a trench rescue IMS system.

- **Technician Level:** This level involves the additional training associated with intersecting and deep wall trench rescue operations. Personnel at this level are primarily responsible for overall operations, development of protective systems, access, and disentanglement of victims in trench or excavation collapses. These personnel are typically technical rescue team personnel who have advanced training above the operations level. Additionally, they will have technical rescue "skill package" training. Examples of this training include rope rescue, confined space rescue, and structural collapse rescue.

- **Instructor level:** This level represents the closure of the training cycle. Personnel at this level are certified instructors who provide the required training to all levels within the organization. At this level the individual should be an instructor and certified in course development. Instructors should also be active rescue team members in order to maintain their skills in a particular area(s) of expertise.

This manual has been designed to provide the student with a building block approach to trench rescue. Therefore, if you start at Unit 1 and proceed through the manual you will have covered all of the information required to satisfy the 1999 Edition of NFPA 1670 Chapter 9 requirements for Trench and Excavation Search and Rescue.

The following is a breakdown of the units in this manual by the Awareness, Operations, and Technician levels should you choose to teach or learn the information to satisfy a particular operational level of the 1999 Edition of NFPA 1670 Chapter 9 Trench and Excavation Search and Rescue. Note: Chapter 9 of NFPA 1670 contains additional confined space and hazardous materials training requirements. See Appendix Five for the 1999 Edition of NFPA 1670 Chapter 9, Trench and Excavation Search and Rescue.

Trench Awareness

Introduction   Trench Rescue
Unit One   Considerations for Specialized Operations
Unit Three   Preparing the Rescue System
Unit Four   Introduction to Trench Rescue
Unit Six   Soil Physics
Unit Seven   Conditions and Factors that Lead to Collapse
Unit Eight   Types of Trench Collapses
Unit Ten   Personal Protective Equipment
Unit Eleven   Equipment and Tools for Trench Rescue Operations
Unit Thirteen   Trench Rescue Assessment
Unit Fourteen   Hazard Control

Trench Operations

Unit Two   Trench Rescue Decision Making
Unit Five   Trench Incident Management and Support Operations
Unit Nine   Soil Classification and Testing
Unit Twelve   Air Bags for Trench Rescue
Unit Sixteen   Gaining Access
Unit Seventeen   Protective Systems in Trench Operations
Unit Eighteen   Victim Packaging and Termination Procedures
Unit Nineteen   Techniques for Trench Protection (non intersecting trenches)
Appendix One   Definitions
Appendix Four   Sample Trench Rescue Tactical Worksheets

Trench Technician

Unit Fifteen   Atmospheric Monitoring for Trench Rescues
Unit Nineteen   Techniques for Trench Protection (intersecting trenches)

Summary

The "big three" refers to special people, special equipment, and special training. All three of these "special" elements work together as part of the specialized training cycle for trench rescue. Depending on the level of commitment in your organization, personnel can be trained to the Awareness, Operations, or Technician levels of trench rescue. While not a recognized level associated with NFPA 1670, the Instructor is widely recognized as the highest level of competence in rescue operations.

Questions (Answers and Discussion on page 237)

1. The term big three refers to specialized rescue operations that require:

   a. Special people, special equipment, and special assignments
   b. Special equipment, special apparatus, and special assignments
   c. Special people, special PPE, and training
   d. Special people, special equipment, and special training

2. The service level(s) as identified by NFPA 1670 that apply to rescue operations and may be an indicator of level of competency are Instructor, Awareness, Operations, and Technician.

   a. True
   b. False

3. Operations level personnel are trained to:

   a. A level where they are only responsible for hazard identification
   b. Provide stabilization efforts in intersecting trenches
   c. The first level at which personnel learn the necessary techniques to work in trenches
   d. Provide the rescue effort advanced skills like rope and confined space rescue

UNIT TWO

TRENCH RESCUE DECISION MAKING

TERMINAL OBJECTIVES

To understand the importance of performing a proper risk/benefit analysis at the scene of every specialized rescue emergency, and to understand the various reasons that specialized rescue operations fail.

ENABLING OBJECTIVES

The student will be able to:

- Describe the theory of risk/benefit as it applies to trench rescue

- Determine the difference between a rescue and a recovery

- Understand the concept of head vs. heart decision making

- Explain the F.A.I.L.U.R.E. acronym as it applies to specialized rescue operations

Risk/Benefit

To fairly evaluate the importance of risk/benefit one must have first suffered through the devastating losses incurred by not using it. As professionals in the field of emergency service delivery, we are most often left "pondering" the thoughts that were going through someone's head at the time of a tragic mistake or critical error. Frequently the conversation revolves around "what in the world caused him to do that?" In all fairness, it is always easier to surmise what you might have done, especially after the results of the decision are known. Nonetheless, our objective should be not to get in that position in the first place.

Of the many factors we can examine that reduce firefighter deaths and injuries, the most prevalent one is always training. The type of training that deals not with tactics, but, instead, with the internal process we use to decide how much risk we are going to assume while acting in the performance of our duties. We call this internal evaluation process "Risk/Benefit Analysis," and although you might not fully understand the principles of risk/benefit, you unconsciously do it each day without realizing the impact it has on your efforts. When dealing with trench rescues, your struggle will be to focus on bringing risk/benefit analysis to the forefront of your strategic decision-making process.

Try to imagine standing beside a trench with a rescue tool called a "risk/benefit scale." This is a set of scales similar to the scales of justice, weighing "risk" on one side and "benefit" on the other side. The tool works by weighing all factors that deal with risk and comparing them with the factors that determine benefit. Obviously, we can turn the situation in our favor if the benefit side is much heavier than the risk. If the risk side seems heavier, there may be no advantage to continuing the operation.

Is this a rescue or recovery?

Since we can never fully eliminate all the risk associated with performing trench rescues, it is vitally important to understanding the difference between a rescue and recovery. You will have committed a terrible disservice to yourself and your team members if you kill or injure them during a recovery operation. Simply stated, it will be your job to evaluate each collapse situation and determine a victim survivability profile. If we are dealing with dead people, does it really matter how long it takes us to get them out? I don't think so! Remember, most dead people stay dead-dead-dead.

What is the risk to the rescuer?

Taking for granted we have addressed the previous points, you now have to turn your attention toward determining the risk facing the rescuers. With all considerations under scrutiny, do they stand a fair chance of succeeding without getting killed or injured? In addition, is the risk to your rescuers proportional to the potential benefit of the attempted action? If you are questioning your judgment at this point in the evaluation, you can be sure you are about to make a big mistake by proceeding with the rescue effort.

Partially trapped worker with good survivability profile

What is the benefit to the situation?

If you can reduce the risk to the rescuer, and the benefit is a savable victim, you are close to giving the situation the green light. No matter what anyone tells you, there is no benefit to saving a dead person, or a dead person's property, if that action requires risk to your personnel. That is the purpose of insurance.

Completely buried victims are usually dead when you get there and just as dead when you leave

Head vs. heart decision making

Remember that compassion kills. In every situation ask yourself, "am I thinking with my head or my heart?" Hoping that you can effect a rescue in light of the fact that the situation is hopeless is thinking with your heart. As much as you might wish that a victim under ten feet of soil in a trench collapse will be alive, without a protective mechanism in place he/she is going to be dead. Nothing you do in that situation will reverse the misfortune of that victim. Just make sure you do not add to the problem by placing your personnel in jeopardy.

If you take nothing else away from this trench manual but the ability to act responsibly regarding your own and others' safety during trench accident mitigation, then I have accomplished my ultimate goal. As the rescuer, do not become the second victim.

THE F.A.I.L.U.R.E. ACRONYM

During the vast majority of emergency events that "go bad" or have major components go wrong, it is possible to identify specific aspects of the operation that contributed to the occurrence. Sometimes these failures result in death or injury to rescue personnel or additional complications in the management of the rescue scene. When dealing with technical rescue operations, the F.A.I.L.U.R.E. acronym is used to describe this process.

- **F: Failure to understand, or underestimating the environment:** In many instances, personnel simply lack the education (awareness) to make the proper decision based on the environment in which they are required to work. Ask any firefighter if he would wear a shower curtain as personal protective equipment (PPE) at a structural fire and he will give you a thousand reasons why this would not be a good idea. Ask any EMS provider why he would not handle a bloody HIV contaminated patient without proper PPE and he or she will provide the right answer.

  So why is it that personnel routinely jump into open trenches without taking the necessary steps to make it safe? Why is it that they simply **fail to recognize the hazards of the environment they are in?** Lack of education, hero syndrome, heart vs. head decision making? There are a thousand excuses, all of them unacceptable. Environmental factors that may not be considered include, but are not limited to:

  - Weight of soil
  - Instability of the trench after the primary collapse
  - Kinetic energy in wall movement (45 mph in some instances)
  - Atmospheric conditions

- **A: Additional medical implications not considered:** The reason we conduct rescue operations is to rescue and take care of patients. If we fail to provide adequate patient care to the victim of a trench collapse, we may end up with a fatality, either immediately or after disentanglement and removal. There are specific medical needs of a trench collapse victim, such as Crush Syndrome.

- **I: Inadequate rescue skills:** You would not go sky diving without the proper training and equipment. So why is it that fire and rescue personnel feel they just have to do something, right or wrong? If you as a responder do not know your limitations and do not have the skills to perform at a certain level, then doing nothing is better than doing something wrong. You should not let your ego cause you to attempt rescue operations for which you are not trained.

- **L: Lack of team work and experience:** Team work is not "many people doing what I say!" Personnel who expect to integrate into trench rescue operations must work together effectively if they are to be successful. This comes from understanding "team decision training," as opposed to simply training and experience. It is not something that just happens because a group of personnel has been trained. Teams are living, breathing entities that have specific identities and capabilities. Individuals do not do as well as collective teams. Team integrity, team processes, and team efficiency concepts are all integral parts of being able to function at the **"high performance end, and not the dysfunctional end of the team scale."**

- **U: Underestimating the logistical needs of the operation:** From day one we have been taught, "Call for it. If you do not need it, turn it around." Now, add to that all of the external resources you may need, the need for trained teams, and the need for special resources that are not used that often and look out! Territorial egos also get in the way of good sense, to the point that we lose our focus on the customer. Nobody likes to be the logistics officer, it's not glorious, but just let something not be where it is supposed to be and see who catches hell! This is not a weekend project at your home where you can wait until Monday to go to the hardware store. You either get it, or it affects your operation.

- **R: Rescue-recovery mode not considered:** Let's face it, most "DEAD PEOPLE STAY DEAD, DEAD, DEAD." Do not commit resources into a questionable situation to recover a body. You have all the time in the world. Very few completely buried victims will survive. Do not let compassion override your good sense. Compassion has no place in the special operations environment where the event is occurring.

- **E: Equipment not mastered:** How well do you think a civilian would do if you were to completely disassemble your SCBA or Life Pak and then tell them to use it in a high stress, emergency operation having never seen it before? Well, the equipment you will use in rescue operations falls under the same parameters. Most of it is specifically designed for the trench rescue operations, and it requires you as the rescuer to know it inside and out. **REMEMBER THE BIG 3?**

TIP: Improve the abilities of the student learning specialized equipment by following this system:

1. Explain and demonstrate the equipment or tool
2. Have the student explain and demonstrate the equipment or tool to you
3. Have them do it again in slow speed
4. Start timing them in decreasing increments to raise anxiety
5. When they become proficient by time, start taking away their sensory perception:

Gloves	Feeling
Blindfold	Sight
Noise	Hearing

6. When they become proficient with gloves while blindfolded and with noise, start timing the evolution again

Now that's creating a "top gun" trench rescuer.

Summary

The theory of risk/benefit deals with the internal process we use to decide how much risk we are going to assume in a given situation. The risk is generally evaluated against the benefit our participation will bring to the event.

In many cases rescue personnel do not consider whether the situation is a rescue or a recovery. In other cases our very nature as rescue personnel leads us to think with our hearts, hoping someone is alive, instead of with our heads when we know the situation is hopeless.

A close examination of rescue situations that go bad can more than likely be attributed to one or more parts of the F.A.I.L.U.R.E. acronym. The bottom line is knowing your environment, considering all medical implications, maintaining great team skill, mandating team work, recognizing the difference between a rescue and recovery, calling for necessary equipment early in the event, and knowing that your equipment will give your rescue the greatest chance of success.

Questions (Answers and Discussion on page 238)

1. The primary factor used in performing a risk/benefit analysis:

    a. Is the victim's family on the scene at the time of the accident?
    b. Does the contractor have the necessary equipment to excavate the victim with rescuers entering the trench?
    c. Is the event a rescue or recovery
    d. How long it will take to remove the victim if OSHA is called for expert help

2. The "L" in the failure acronym describes:

    a. Lack of rescue skills
    b. Lack of equipment
    c. Lack of medical knowledge
    d. Lack of teamwork and experience

3. Most completely buried victims ultimately survive a trench collapse.

    a. True
    b. False

UNIT THREE

PREPARING THE RESCUE SYSTEM

TERMINAL OBJECTIVE

To understand the various methods that can be used to form and maintain a fully functional and effective team for trench rescue.

ENABLING OBJECTIVES

The student will be able to:

- Discuss the advantages and disadvantages of being a self-sufficient, community dependent, or regional trench rescue team

- Determine the most advantageous physical and mental characteristics of potential trench team members

- Explain the T.E.A.M. acronym as it applies to trench rescue

- Describe the weight, size, and characteristics of the equipment used in trench rescue

- Understand the need for choosing the most appropriate method to move and store trench rescue equipment

- Explain the advantages and disadvantages of each type of trench apparatus

If you are considering developing a rescue system for trench emergencies, you can pursue a number of different avenues. Each one has pitfalls and consequences in the form of time, money, training, and equipment. The road you choose should reflect the expectations of your community for that type of service, and your organization's level of commitment toward providing the necessary resources to handle emergency events safely.

The most difficult and time-consuming method to provide trench rescue to your community is to be **self-sufficient**. This means that your agency has committed to many hours of training and practice for your personnel. In order to effectively train, you will need considerable specialized equipment and also a means of transporting this equipment to emergency scenes. The bottom line...this is, no holds barred, the best method to ensure your community is prepared for a trench emergency. But at the same time, it is the most expensive. Remember the "Big Three."

Phone and Resource List

Trench Resource List		Contact #
Name	Item	
Splinter's Lumber yard	Panels, bracing	Joe (202) 555-1767
Gopher's Construction	Crane, back hoe, ventilation fan, ladders	Bob (202) 555-3347
Percy's Pumps	Wet/dry pump; Air bags, pump	Percy (202) 555-2578

Many areas recognize the expense involved in self-sufficiency and elect to be **community dependent**. This may be for a number of reasons, including the infrequency of previous trench collapses, or because a trench collapse has never occurred. If you decide to be community dependent, seek out those in your area that have construction and excavation experience and put them on a "call back" list. It is also a good idea to predetermine necessary logistical needs and supply a list to the specific vendors before an emergency occurs. For instance, your local hardware store may be more than willing to keep a cache of lumber and nails in a specific place for your immediate use should a trench collapse happen. Being community dependent could possibly address logistical needs; however, your personnel will still need to be adequately trained.

The most cost-effective method of providing trench rescue services to your community is the **regional approach**. The key to this type of service delivery system is that it spreads the cost of development and operation over several jurisdictions.

Community dependent trench vehicle for a department on a budget

This reduces individual organizational costs, but necessitates written mutual aid agreements to be effective. There also has to be a commitment on the part of each participating organization to train and keep their personnel proficient at trench rescue operations. This type of system is very hard to organize and maintain because equipment and team members could be located in many jurisdictions.

YOUR TEAM

If you have decided on the appropriate service delivery system, and have committed to providing trench rescue for your community, your next step will be to put together a team. This is a very important step, because the key to successful trench rescue operations will be more dependent upon the makeup of your team than on the tools with which they operate.

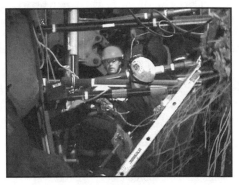
Trench rescues are very tough incidents –teamwork is critical to success

Trench rescue activities are very demanding, and necessitate the use of heavy, often cumbersome equipment. For that reason it is necessary that the personnel on your team be physically fit for the demands of conducting a long-term operation. Moving panels and digging dirt is hard work. Make sure the people on your trench rescue team are up to the demands of such work.

In addition, your team members need to be mentally fit. That is a tough call because who is to say any of us in this line of work are mentally fit. Regardless, search for those personnel who work effectively under stress and when things are not going well. If they are capable of working around a half-buried dead person, or a screaming, combative patient with two broken femurs, they are probably good candidates for a trench rescue team.

Having folks with the proper skills in the proper place is critical to your success and efficiency

One of the most important abilities of a team member involved in trench rescue operations is construction skill. Let's be honest, at some point someone is going to have to cut a piece of wood or hammer a nail. Don't take a 100-pound rescuer, put a 22-ounce hammer in his/

her hand and expect miracles. Team proficiency in construction skills will make your operations safer, more efficient, and less time consuming.

Another important part of the team rescue package is the medical personnel. There is a real need for you to train your medical personnel for the type(s) of rescue in which they will be involved. Special preparations, such as having protocols for Crush Syndrome, may be vital to your victim's survival. In addition, treating a patient in a partially collapsed trench can be a taxing experience. Wouldn't it be better to train your medical personnel to operate in such environments? With such training, they can concentrate on patient care instead of their own well being at an actual incident.

Not to be overlooked is the ability of your personnel to "think on their feet." Taking individual practical training components and applying them in a variety of unique situations is normally required during a trench rescue. The things you read in books and the trenches that you train in will look nothing like your first collapse scene. Make sure you take some thinkers with you!

Recognizing that everyone does some things better than others will also be key to your success on a collapse situation. Put a hammer in my hands and one of two things is going to happen: I get hurt or I hurt somebody else. Is this also a description of your capabilities? If so, then your talents may instead be that you are the best organizer in the world and a great incident command officer. The point is placing the right person in the right job. If you can hammer, you should have a hammer in your hand. If the shovel is smarter than you are, stick to digging!

THE T.E.A.M. ACRONYM

The most important attribute of your team members will be their ability as team players. Not everyone on the scene is going to get a glory job. Nor will every suggestion given to the incident commander be implemented. Likewise, some people will be telling people what to do, and others will be required to follow directions. The duty of a team player is to work as hard as possible to achieve a successful outcome in a given situation, regardless of who made the decision (or how wrong you may think the decision is).

The team-orientated rescuer can take direction and give it, and he/she will flourish at whatever job you give them. It doesn't matter to the team player whether he has to hammer, dig, or direct. The outcome of the rescue is the most important consideration. The thing to remember is that the most talented person might not be the best team player. You will almost inevitably be better off with the latter.

The T.E.A.M. acronym is used to describe the essence of the team philosophy. Each member of the team understands that acting individually is not as effective as working as a team. Remember: **T**ogether - **E**veryone - **A**ccomplishes - **M**ore.

As a final thought in this section on teams, before you make another decision, call those who have done it before. Don't make the mistakes that someone else has already made. Be smart. Look for a successful "up and running" team and follow their example.

GETTING YOUR EQUIPMENT TO THE SCENE

There are any number of ways you can carry and store your trench rescue equipment; however, keep in mind that we are for the most part not talking about sophisticated rescue equipment. We are talking about shores, panels, chains, shovels, buckets, etc. You get the picture. It is construction equipment that is bulky and heavy. The nice thing is that the possibilities for moving the equipment are endless, because the truck doesn't have to be a fancy rescue rig. Although if you have money, well...!

The one type of truck that isn't appropriate for moving your trench equipment is the prototypical vehicle extrication or **squad truck.** These trucks are designed with outside compartments that are too small for the storage of trench rescue equipment. Additionally, the inside is closed and narrow and better designed for people transportation than panel transportation.

A dump truck vehicle can be used to transport panels and other equipment

For this reason, many teams will resort to utilization of a **dump truck** or **flat bed vehicle** that they obtain from their city surplus department. The back of the unit is big enough, and the chassis strong enough, to handle the type of weight represented by shores and panels. The advantage to this arrangement is that the unit is self-contained, relatively cheap, and can sit outside covered with a tarp when not in use.

There are many different **utility trailer** configurations that can be purchased for trench purposes. As long as the wheelbase and chassis will support the load you will be fine. If the trailer is large enough to hold the panels it makes little difference if it is closed or open. The disadvantage is that another motor-powered vehicle is needed to get it to the scene, and it can be more difficult to maneuver. The advantages include low cost and low maintenance.

Custom and converted vehicles are usually designed and purchased by teams that will supply more than one technical rescue service. For instance, the Virginia Beach Fire Department took an old Seagrave pumper and put a Hackney body with roll-up doors on the chassis. They then purchased a trailer to carry the big stuff. By split-

VBFD Tech unit & trailer

ting available storage space, they are afforded the opportunity to provide confined space, rope, and vehicle rescue with the engine, and trench and structural collapse with the trailer. This allows them to leave the trailer at the station when

Pod type vehicle

it is not needed. The disadvantage to this arrangement is that the truck and trailer collectively are sixty-three feet long.

Another type of system that has gained popularity over the years is the **pod** system. Utilizing this approach, a pod equipment storage system is designated for a specific purpose and only that pod is taken to the scene. The truck just backs up and takes whichever pod is appropriate.

TIP: Make sure you figure the total weight of the equipment before developing the specifications for the chassis and axle(s) on your apparatus. When totaled, I think you would be surprised at the weight. A single homemade trench panel can weigh 200 pounds! Also, make sure it fits in the station bay!

Getting folks to the scene on a budget

Summary

There are basically three variations of how a team can be designed and operate. Before you start a trench rescue team, deciding whether to be self-sufficient, community dependent, or regional should be evaluated for their advantages and disadvantages to determine which type is most appropriate for your community and will be supported by the organization.

By far the most important aspect involved in the development and potential success of a trench rescue team is the team members. Look for those folks that can take direction as well as give it. Keep in mind that the best team players are not always your most talented people. The T.E.A.M. acronym, together everyone accomplishes more, is the heart and soul of a good rescue team.

Trench rescue equipment is for the most part bulky and heavy. The type of transportation system you employ to move and store your equipment should take into consideration the weight, accessibility, and maintenance requirements of the equipment.

Questions (Answers and Discussion on page 238)

1. There are a number of different methods to move and store your equipment. Which of the following would be least appropriate:

   a. Dump truck
   b. Flat bed truck
   c. Utility trailer
   d. Vehicle rescue truck

2. The most reliable form of rescue team development is the:

   a. Regional approach
   b. Community approach
   c. Self-sufficient
   d. Multiple team approach

3. By far the most important aspect concerning the potential success of your team will be:

   a. The equipment you have to effect the rescue
   b. The people you choose for your team
   c. The vehicle that transports the equipment to the rescue scene
   d. The commitment on the part of your organization to fund the team

UNIT FOUR

INTRODUCTION TO TRENCH RESCUE

TERMINAL OBJECTIVES

To understand the necessity of maintaining a trained and competent trench rescue team in your community.

ENABLING OBJECTIVES

The student will be able to:

- Describe the conditions that require compliance with the Excavation Standard, and the emergency service organization's relationship with OSHA pertaining to trench collapse operations

- Describe the history of the OSHA Standard on Excavations and explain how the current standard is performance based

- Provide an understanding of OSHA's Standard on Excavations, its enforcement role, and subsequent relationship with emergency service organizations

- Explain the reasons for non-compliance with the trench standard based on the cost and installation of traditional sheeting and shoring

- Understand what types of emergencies can occur at a trench site that do not involve a collapse

- Discuss trench injury and fatality statistics as they compare to other areas of construction

- Recall from memory trench terminology as identified in the excavation standard

- Explain how cost and demographics play a role in non-compliance

- Describe ways in which machines and rigging can fail and create emergencies at a trench site

- Discuss the potential problems that can occur from below grade atmospheric hazards

Trench and excavation safety are primarily covered under **OSHA CFR 1926 Subpart P.** The definitions we will discuss come directly from that document and the others outlined in the introduction of this manual.

In general, we dig trenches and excavations for a variety of reasons. Literally hundreds of trenches and excavations are opened daily. They are used for the following reasons:

- Placement of utilities underground: water, natural gas, and other fuel, electricity, sewage and drain systems
- Removal of old utility systems
- Removal of underground storage tanks (UST)
- Foundations of buildings or large high rise operations
- Construction of basements

Utilities in a trench

Workers taking a risk working in an unprotected trench

Statistics show that accidents occurring in trenches have a much higher fatality rate than other construction accidents—about 100 fatalities per year and approximately 1,000 to 1,500 injuries, of which many are permanently disabling. Some of these victims are rescuers or co-workers attempting rescues of initial victims in collapse situations.

For the purpose of trench rescues there are only two types of trench collapses. "Those in which the victim is dead, or is going to die, and those in which we can save the victim." Recall the F.A.I.L.U.R.E. Acronym, "most dead people stay dead, dead, dead!"

As with all emergency incidents, fire and EMS services will be asked to respond to a variety of trench and excavation emergencies. Depending on the educational level and safety programs of contractors in your area, you may find yourself responding to more trench collapses. Also, the frequency of trench collapses depends on the amount and type of new construction projects underway in your jurisdiction. If you have any of the following, you can expect to see trenching activity:

- Housing development construction
- New water, sewer, gas or electrical installations
- Tunnel or major water projects
- Pump stations
- High rise construction

Telephone utilities in trench

Usually, trench collapse operations are busy, confusing events that require extended operations, specialty equipment, and specially trained personnel. First response personnel who are not provided with Awareness level training become part of the problem rather than part of the solution. Couple this with command officers who have almost no experience in managing these types of operations, and the confusion grows exponentially.

OSHA CFR 1926 SUBPART P, EXCAVATIONS

Understanding the excavation standard is important to rescue personnel for several reasons. Primarily, it will give you the data and information needed to decide on appropriate protective systems and safety requirements for trenches. This information can be universally applied to any given rescue operation by using the "tool box" approach. Your "toolbox" should be full of ideas and techniques, most of which are not appropriate for all situations. The moral here is to maintain a large and varied (Tim Allen size) tool box of information. Remember, if all you have is a screwdriver, it looks a lot like a hammer when you have to drive a nail.

Secondly, knowledge of the standard, its requirements, protective systems, and soil classifications will qualify the user as a **"Competent Person"** according to the standard. This provides some liability cushion, but more importantly it allows the user to make rational decisions based on a given standard during rescue operations.

We can evaluate the history of the trench standard as follows:

- Previously part of the "Contract Work Hours Standard Act"

- Contents and requirements to meet the standard were confusing which led to inadvertent non-compliance and insufficient protective systems. Typically, the protective systems used were more expensive to put into place than the fine associated with non-compliance.

The current Standard still retains about 80% of the original act contents; however, it has been clarified to assure that the requirements can be better understood. Chief among the changes and additions to the new standard are the following:

1. All criteria are performance-based standards. This means that protective systems that are not outlined in the appendix may still be used if available data shows such systems are "performance tested and oriented."

2. A consistent soil classification methodology is delineated, including the techniques utilized to test soil samples. This allows protective systems to be designed according to soil profiles.

3. Flexibility with regard to protective systems development is allowed.

4. Fines and penalties have been increased. Many fines are as much as seven times greater than the amounts specified in the original standard and may even provide for equipment seizure and impoundment during investigations.

The Standard is divided into several key areas:

- **Scope and Definitions**
- **General Requirements**
- **Protective Systems**
- **Appendices**

GENERAL REQUIREMENTS

General requirements are those items required during construction operations that a competent person must consider and act upon. From a rescue perspective this offers an excellent "safety guideline" from which we can draw tactical decisions.

Consider the following as **rescue safety and operational guidelines and considerations.**

- **All trenches must be protected before entry <u>except:</u>**

  - Those made entirely of stable rock
  - Those less than five feet in depth, previously inspected by a competent person, and found to have no indication of a potential cave-in

- **Protection:** Any trench more than five feet in depth, including the height of the spoil pile must be protected.

- **Spoil Pile:** Must have a two-foot setback from the lip.

- **Egress:** Trenches four feet or greater in depth must have a means of egress every twenty-five (25) feet. LADDERS!

- **Atmospheric hazards:** Trenches four feet or greater in depth must be tested before entry if an oxygen deficient or other hazardous atmosphere could exist. However, all trenches should be tested for the following:

  - Oxygen deficiency or enrichment (less than 19.5% or greater than 23.5%)
  - Hazardous atmosphere (toxins in PPM)
  - Flammable gases (greater than 10% of the LEL)

  Testing must occur as often as necessary to ensure a safe atmosphere, and emergency rescue equipment must be readily available when a hazardous atmosphere could exist. A trench is an excavation, and therefore exempt from OSHA 1910.146 (the Confined Space Standard). Voluntary compliance with 1910.146 requirements will provide additional life safety and liability protection. The crucial point is that workers/rescuers must be protected by either ventilation or respiratory protection if the potential for an atmospheric hazard exists. See Unit 15 for more information on atmospheric monitoring.

- **Water accumulation:** Employees need to be protected from water by protective systems, dewatering operations, and/or a lifeline and harness where applicable. Remember, dewatering must be monitored by a competent person and surface runoff must be diverted.

- **Soil:** a competent person must be able to determine the soil classification.

- **Inspection:** A competent person must inspect the trench (even during rescue operations) for:

Dewatering operation

  - Secondary cave-in potential
  - Protective systems failure
  - Atmospheric monitoring or control
  - Other hazardous conditions (**can you think of some?)**

TRENCH TERMINOLOGY

A **"trench"** means a narrow excavation (in relation to its length) made below the surface of the ground. In general, the depth is greater than the width, but its width measured at the bottom does not exceed 15 feet.

"Excavation" means any man made cut, cavity, trench, or depression in an earth surface formed by the earth's removal. Again, in practical terms, when a hole is more than 15 feet wide at its base, it is called specifically an excavation. Overall, an excavation is wider than it is deep.

Other terminology that is important at this point in the manual is:

FLOOR: The bottom of the trench.

WALLS: Anything that is in the vertical or upright, on the long axis.

ENDS: The ends of the trench, where the walls end at the short axis.

LIP: The area 360 degrees around the opening of the trench and extending down two feet. Very dangerous.

TOE: The area where the walls and floor intersect at the bottom of the trench and two feet up.

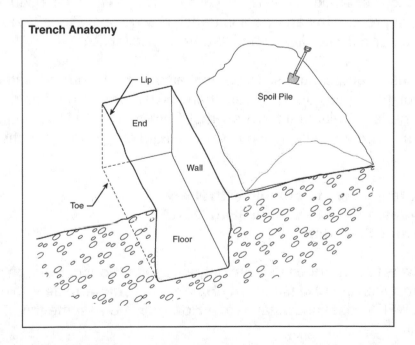

Trench Anatomy

COMPETENT PERSON: The individual, usually the supervisor, who meets the OSHA standard for determining soil profiles, safety concerns, protective mechanisms, and other performance requirements.

All of this terminology, and other related terms, will be used throughout the lesson. If you have a question about a specific term, please refer to the definitions portion of the manual in Appendix One.

OSHA AND TRENCH RESCUE

Fire and rescue agency relationships with OSHA are unique among compliant organizations. Since the OSHA standard for "Excavations" (29 CFR 1926 Subpart P) was created for the construction industry, how we fit into the picture is usually up to a local OSHA Enforcement Officer. The standard may be applied quite differently from one jurisdiction to another. For rescue operations, it simply makes good sense to locate a standard that has been adequately researched, whether it is OSHA, NFPA, or another standard, and use it to our advantage.

From a rescue agency standpoint, OSHA requires compliance with the standard if the following conditions exist:

• An employee/employer relationship exists

• Trench rescue operations are part of your job. Do you train for, acquire equipment for, or prepare to participate in trench rescue operations? Additionally, would you normally be dispatched to a trench collapse?

Paid versus volunteer status issues may also arise. If you are paid and can answer any of the above questions yes, then you will probably be held accountable under the law. Volunteers are exempt from most OSHA regulations, so CFR 1926 may not apply in a given rescue situation. However, consider the following:

• **Do volunteers have to comply with NFPA?**
• **Do volunteers have to comply with OSHA?**
• **Are you an OSHA state?**

The answers to these questions are only as clear as your jurisdiction chooses to make them. Often the **AHJ (Authority Having Jurisdiction)** determines compliance with NFPA consensus standards. However, we assume that if you are reading this manual to train your folks, you are determined to give your personnel the safest, most efficient environment to work in and, therefore, would not ignore the standards.

Whether you agree with this, or any other legal or consensus standard, should something go wrong you can bet it will end up in court. Then someone (an expert) will show up in court quoting the law and waving nationally recognized **consensus standards**. In effect, using them to the prosecution's/plaintiff's advantage. **Get the point?**

Consider the following when talking about the liability aspects of trench rescue operations:

- What is the current "standard of care" for a trench rescue?
- What guidelines do similar teams follow?
- Do you comply with other OSHA or NFPA standards?

HOW OSHA VIEWS TRENCH RESCUE OPERATIONS

During trench rescue operations we tend to "over-engineer" our rescue systems to protect ourselves from the worst case scenario. In effect, we usually build our systems because we understand that the most important people on the scene are not the victims, but the rescuers.

Additionally, the time we are in the trench is limited and, therefore, considered by OSHA in compliant specific situations. Usually our time in the trench is limited to hours, not days. The OSHA standard is designed to regulate protection systems engineered to last many days.

Finally, we are entering the environment for an entirely different reason than a utility worker. Rescuers are entering to do rescue operations and not commercial construction.

This is not to suggest that we should be surprised or embarrassed if OSHA decides to get involved. Remember, the field belongs to them, even if we are playing the game. Generally, OSHA will get involved if one of the following occurs:

On a real trench collapse panels and shores often don't line up—still we tend to over-engineer our protective systems

- There is a civilian or rescuer injury or death as a result of the collapse. (In fact, you should call OSHA whenever you have a civilian/rescuer injury or death.)

- The death of anyone as a result of a construction incident

- When requested by the Authority Having Jurisdiction

In reality, the issue is one of cooperating with OSHA. We should not view OSHA as the enemy. We should keep them abreast of our activities in a rescue situation. Also consider taking time before the emergency to meet and confer with your local OSHA representative. Together, you should work toward an understanding of each other's roles. Often they can provide a wealth of information and site support.

NONCOMPLIANCE AND TRENCH COLLAPSE EMERGENCIES

The chances these workers are taking happen in every community

As a contractor preparing to lay a pipe string, I have some decisions to make. Obviously, I should determine the soil profile and use a pre-engineered system based on the soil's potential for collapse. The choices I have are as wide and varied as the different types of materials that can be used to construct the system. Steel interlocking panels, pre-formed steel or aluminum trench boxes, solid wood uprights, and/or hydraulic or air shores are just a few of my choices. More on these types of systems later.

The one consistent factor a contractor would closely evaluate regarding the type of trench protection utilized is the cost in time and money for that system. It takes a long time to panel a ten foot long, four foot deep trench, just to lay one section of pipe. Why not take a chance, dig the trench, lay the pipe, send someone down there to make the connection. You get the picture. The faster I lay the pipe, the more money I will make, and ultimately the faster I can get to another job site.

Commercial sheeting and shoring trench operation

Workers in an unprotected trench

Another issue that no one is comfortable addressing is the socio-economic makeup of the trench collapse victim. I don't know about you, but I have not rescued very many "three piece suits" from trenches, which means the guy in the trench is probably making minimum wage at best and, additionally, needs to keep his job to support a family. Is this individual going to question the safety of a job site or trench? The sad part of our job as rescuers is knowing that the victim does not normally understand the hazards involved, and if he did, he would not be in a position to question them.

ACCIDENTS WITHOUT A CAVE-IN

A collapse is not the only type of emergency that you will respond to involving trenches. In fact, when it is all said and done, most of the emergencies in trenches deal with an occurrence other than a collapse.

The plain fact is that a lot of work goes on after the trench is dug. This work takes place in protected and non-protected trenches. The challenge for the rescuer is to not be lulled into complacency by protected trench situations, or to assume that these rescues will be easily accomplished.

EQUIPMENT FAILURE AND LOAD MANAGEMENT

One of the problems you will be confronted with at the scene of a trench emergency that doesn't involve a collapse is the dreaded backhoe or excavator caused problem. Being mechanical, and subject to malfunction, these machines can cause terrible problems for workers operating on and around this equipment. Hydraulic failures during lifting operations and rigging that is improperly assembled or not appropriate for the load being moved result in numerous construction accidents.

Accident without a cave-in—backhoe in trench

Keep in mind all of the things that have to go right with just the equipment at a work site. The backhoe has to work properly, the load has to be rigged properly, and the rigging has to be substantial enough to carry the load. When problems occur in any one of these areas, you may well find yourself in a trench rescuing a trapped or pinned worker.

The machines operating at a trench site are powerful and the loads being lifted are often heavy. Workers are frequently pinned between steel panels they are trying to set as sheeting, or by pipe that is being placed. When this happens, you as the rescuer may be faced with a trench that is only partially protected, and a seriously injured victim.

To make matters worse, the backhoe or excavator is operated by humans in situations where tolerances for maneuvering the load are small. Water and sewage pipes, as well as steel plates being used as sheeting panels, have been known to crush workers while being moved into place. Commonly, these scenes call for a rapid sizing up of the protective system being used and continuous evaluation of that system to determine if your extrication methods compromise the in-place system.

RIGGING

Another problem is that the loads being moved in and around trenches are only as safe as the rigging and rigger that secures them. If a rigging strap breaks or a crane's hydraulic system fails and a pipe falls on a worker, you could be faced with a "Mr. Pancake Man." Even if the trench is protected, getting the victim out may be a huge challenge.

ATMOSPHERIC CONCERNS

Atmospheric problems are another area of concern, and a frequent cause of problems at trench sites. With today's stringent Haz-Mat laws, it is not unusual to

find hazardous waste products buried underground. If you happen to be a worker in or around a trench when one of these containers is broken, you could be confronted with an atmosphere that is well within the explosive ranges, above permissible exposure limits for toxic atmospheres, or accompanied by a low oxygen profile.

Since it is impossible to determine what someone may have previously buried, extreme caution should be used when arriving at the scene with workers down in a trench. If there is just one worker involved, one might assume an injury or a medical problem exists. If two or more victims are involved, some internal alarm should be sounding. In law enforcement they call that a "clue!"

Monitoring must be done at all levels of the trench

If you are confronted with an incapacitated worker in a trench for no apparent reason, remember what we said earlier about the trench environment and how critical it is to monitor the atmosphere. The rule of thumb in this situation is that if you have one worker down it may have been a heart attack or some other illness; if two or more workers are down, and no accident is apparent, you are most likely dealing with a hazardous atmosphere. Remember to monitor. Also have a Haz-Mat Team available as a part of your initial response to any trench collapse.

Summary

Because trenches have a higher fatality rate than other areas of construction, the Federal Government established OSHA CFR 1926 Subpart P on Excavations. The original excavation standard was called the Contract Work Hours Standard Act.

The general requirements contained in the excavation standard should be considered safety guidelines for trench rescue and include:

* Protecting all trenches five feet in depth, including the spoil pile, except stable rock and those inspected by a competent person and found with no potential for collapse
* Setting the spoil pile back a minimum of two feet from the trench lip
* Providing a ladder or other escape method every twenty-five feet in a trench
* Always monitoring for a hazardous atmosphere

Technically speaking, any hole cut in the earth could be deemed an excavation; however, for trench rescue purposes, a trench is generally deeper than it is wide but never greater than fifteen feet wide at its base. Other important trench terminology includes floor, walls, ends, lip, toe, and competent person.

Complying with OSHA standards as a rescue team, paid or volunteer, is a tricky legal question for many jurisdictions. As competent professionals, we should use NFPA and OSHA standards because they are created with the best interest of our personnel in mind and adopted appropriately for that purpose.

There are many causes of trench accidents. From a collapse perspective most of the failures have to do with non-compliance on the part of the contractor in order to complete the job faster and make more money. However, many of the accidents in trenches have nothing to do with the collapse of the trench itself but rather human, mechanical, load management, or atmospheric reasons.

Questions (Answers and Discussion on page 239)

1. Statistics show that trench accidents have a _____ fatality rate than other types of construction accidents:

   a. Higher
   b. Significantly lower
   c. About the same
   d. Unknown

2. The OSHA standard for trenches and excavations is:

   a. OSHA CFR 1926 Subpart P
   b. OSHA CFR 1910.146 Subpart P
   c. OSHA CFR 1910.126 Subpart C
   d. OSHA CFR 1910.123 Subpart P

3. On the scene of a trench or excavation you could expect to find a person who is familiar with all aspects of soil types and testing called:

   a. Knowledgeable person
   b. Testing and compliance officer
   c. Expert excavation and testing inspector
   d. Competent person

4. A trench is an excavation that is generally deeper than it is wide but its width measured at the bottom does not exceed 15 feet.

   a. True
   b. False

5. Egress ladders in a trench must be within ___ feet of a worker in a protected trench:

   a. 10
   b. 20
   c. 15
   d. 25

6. The overriding reasons for contractor non-compliance with trench protective systems is:

    a. Time and money considerations
    b. Equipment and manpower requirements
    c. Testing and competent person requirements
    d. Lack of training and education

7. The minimum setback requirements for the excavated spoil pile is:

    a. 4 feet
    b. 2 feet
    c. 6 feet
    d. 12 feet

8. The OSHA standard on trenches and excavations is :

    a. An analytical based standard
    b. A performance based standard
    c. A preliminary study standard
    d. A construction industry standard
    e. Both b and d

9. The OSHA standard was originally part of the:

    a. Contract Work Hours Standard Act
    b. Standard Act for Workers who make Minimum Wage
    c. Standard for Construction Workers Act
    d. Contract Employees Act for Competent Persons

10. The height of the spoil pile is taken into account when determining the need for a protective system:

    a. True
    b. False

UNIT FIVE

TRENCH RESCUE INCIDENT MANAGEMENT & SUPPORT OPERATIONS

TERMINAL OBJECTIVE

To understand the importance of the Incident Management System (IMS), and its various support functions, as applied to successful trench rescue operations.

ENABLING OBJECTIVES

The student will be able to:

- Explain the various components of an Incident Management System for trench emergencies

- Describe the various IMS support functions and their importance to successful trench operations

Incident management for trench collapse emergencies is not unlike any other IMS used for fire or vehicle accidents. There will always be strategic, tactical, and task levels. Someone will need to be in charge, and others will need to follow directions. Having a clearly defined approach to incident scene responsibilities and authority is critical to the safety of your patient and rescuers.

Dividing trench scene duties and responsibilities allows the Incident Commander to implement a systematic method to handle a problem that can quickly overwhelm even the most effective and experienced officers. It also decreases the organizational span of control, and provides a measure of on-scene accountability.

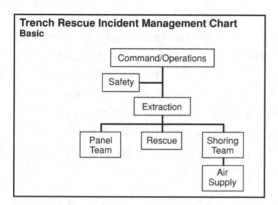

Ultimately, the level to which you develop your IMS is going to be entirely dependent on the magnitude of the problem, and the number of resources you will have on the scene. If, for instance, your rescue team is comprised of yourself and three others, it makes no sense to try to fill all of the job functions. For obvious reasons, each person will fulfill multiple roles. In effect, you are the system.

Listed above are the typical areas of the command structure that will need to be filled at a trench collapse emergency.

The Strategic Level

The **Incident Commander** is responsible for developing the strategic goals for the operation. This person is ultimately responsible for determining and arranging the acquisition of all resources necessary to handle the incident. For instance, the Incident Commander may develop a strategy specifying that the victim of a major collapse will be recovered using commercial techniques as provided by John Doe construction company. He would then call for the resources necessary to ac-

Incident Command

tually fulfill the strategic goals. How this will exactly take place will rest with the Operations Officer. More on the Operations Officer later.

Directly under the Incident Commander is the **Safety Officer**. This function needs to be filled with someone who cannot only spot unsafe acts, but who also has the ability to anticipate activities that may lead to an accident. While everyone on the scene is responsible for his or her own safety, it is necessary for someone to control the "big picture." It is critical that the Safety Officer be familiar with the environment and its potential hazards. This is a very important position when addressing a trench emergency, and consequently is the only individual other than the Incident Commander that can halt the rescue effort for any reason, at any time.

The **Liaison Officer** position will often have to be assigned on a trench collapse operation. This is because you can expect that many agencies other than your own will be involved in the rescue effort. Someone has to gather critical inter-agency information from all parties involved, and at the same time buffer the Incident Commander from being overwhelmed with information that is not critical to the decision making process. Some examples of these other agencies include the Police Department, Utility Contractor, Electric Company, Water Department, Red Cross, and OSHA, just to name a few.

The **Public Information Officer**, if established, will provide the media with a direct point of contact for on-scene information. This person can provide frequent updates to the media on the progress of the rescue effort and, more importantly, educate them regarding your rescue methods and difficulties. Keep in mind, these events don't happen every day. For that reason you will need to get as much positive coverage out of each occurrence as possible. Picture the PIO on your first collapse telling the media that the rescuers are doing all they can, but that they lack the specialized equipment necessary to speed up the operation. Next thing you know, the equipment purchase requisitions are flying out the door like they're on fire. To me, that's good stuff.

The Tactical Level

The tactical level of the incident management structure takes the strategic plan developed by the Incident Commander and implements the tactics necessary to achieve success. Using the previous example, this level would decide on the actual commercial techniques that would be required. The functions at this level consist of the Operations, Logistics, Finance, and Planning sections.

NOTE: Since trench emergencies rarely necessitate the use of a Finance or Planning Officer we will leave the explanation of those functions for a class on IMS, and instead concentrate on those areas that are almost always filled.

The **Operations Officer** is the person who actually runs the incident. The individual assigned this position is responsible for overall coordination of the rescue effort, and the implementation of the tactical decisions that will make the Incident Commander's strategy successful. Included in the Operations Officer's chain of responsibility are all of the individual groups that provide direct emergency support to the trench rescue effort, including the Extrications Officer, Panel Team, Cutting Team, and Shoring Team.

The **Logistics Officer** is responsible for obtaining the appropriate equipment and personnel for deployment by the Operations Officer.

Task Level

The **Medical Officer** normally works at the direction of the Operations Officer and is responsible for establishing a medical control area to treat any on-scene rescuer injury, and to provide for patient care as is necessary. He may also establish a transportation section in the case of mass casualty disasters. Additional duties may include assisting the Rehab section with monitoring of rescuer vital signs.

The **Extrication Officer** is responsible for the actual extrication of the victim and for all of those activities that are required to facilitate the rescue. This position coordinates the various support functions and ensures the proper steps in the recovery effort are followed.

Emergency Support Functions

Not all personnel at a trench collapse are going to get glory jobs. Nor do all people want them. However, the trench support activities are a very integral part of the trench rescue operation and cannot be neglected if you are going to be successful. It takes a tremendous amount of manpower to handle a trench collapse situation, and most of it will be performing auxiliary functions. Below is a list and description of some of these support functions.

Air **supply operations** will be necessary if you are considering using pneumatic air shores or air bags. Persons assigned this task will need to assure proper operation of the equipment and gather and secure the necessary air supply.

The **cutting team** will be responsible for all cutting and manufacturing of systems that contain wood. Examples of cutting jobs include wedges, strongbacks, and shores. For this function make sure you get competent people who can handle a saw.

The **panel team** is required to set up, carry, and install all shields or panels. Make no mistake about it, being on the panel team is hard work. The panel team needs to be a group of at least four people who are your designated "mongos." (A "mongo" is a guy who is bigger and stronger than he is smart, but when you give him a job there is no stopping him.) Keep in mind, their job is one that goes like hell when panels are called for, then has slack time until termination. Consider reassignment of the panel team after inside panels are set.

The **shoring team** will be required to assemble and install all shores and wales required to make the protective system. This may involve the installation of wood, pneumatic, or another type shoring. The persons assigned this role need to have a great deal of manual dexterity and be efficient with hammers, nails, and other hand tools. More on this in the shoring section.

TIP: Remember to always establish a rapid intervention team (RIT) prior to any trench stabilization activity.

Logistics

Equipment and logistical support functions entail all areas of equipment storage and dissemination that takes place on the scene. It is vitally important to keep all equipment not currently in use at a predetermined location. That way, someone can keep track of it and determine its availability at any given time during the emergency.

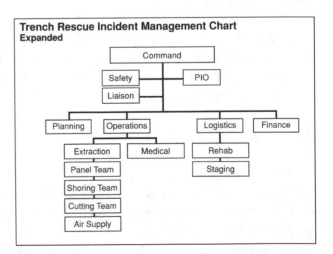

The **Staging Officer** is responsible for ordering and maintaining adequate resources at the scene to handle additional requests for equipment and manpower. This would include establishing an area to stage manpower for future deployment, and to maintain/store equipment.

For trench work, the **rehabilitation sector** is an absolute necessity. Rescuers will be working very hard for long periods of time and will need rest. This function should provide for rescue personnel rotation to address medical monitoring and fluid replacement needs.

> **TIP:** If you are an Incident Commander, make an effort to thank the support people and recognize their contributions. The overall success of the incident will be directly related to success of trench rescue support functions.

Conclusion

I am sure that everyone has heard the old metaphor "a good house starts with a good foundation." Applying this to the trench rescue concept infers that a little time spent planning and organizing your efforts at the beginning will save time and maybe lives at the end. A well-developed and expanded IMS will be crucial to your success at a trench emergency. Just remember, develop a system only as large as necessary to handle the incident.

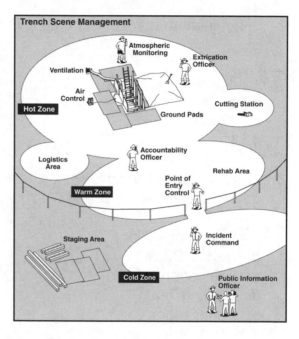

Summary

Trench rescue incident management can be broken down into the strategic, tactical, and task levels of operation. Strategic operations involve the big picture and what needs to be done, while tactical and task levels involve the how we are going to get it done part of the equation.

Several support functions needed at a trench emergency may include but not be limited to panel team, air supply, shoring, cutting team. Additional functions that are very important are the logistics, staging, and rehabilitation sections.

In any rescue operation, personnel should be designated as a Rapid Intervention Team (RIT). The RIT should be properly equipped and assigned the primary function of immediate rescue should someone involved in the trench collapse become trapped or hurt.

Questions (Answers and Discussion on page 241)

1. The operations officer at a trench collapse:

   a. Is responsible for tactical decisions on the rescue site
   b. Should be one of your most experienced technical rescue team officers
   c. Reports directly to command
   d. Supervises the extrication officer
   e. All of the above

2. The minimum number of persons that should be assigned to a panel team is:

   a. 3
   b. 4
   c. 2
   d. 1

3. Which of the following is not a critical IMS position assignment when initially getting started at the scene of an incident?

   a. Operations Officer
   b. Rehab Officer
   c. Extrication Officer
   d. Shoring Team
   e. Panel Team

4. A Rapid Intervention Team (RIT) should always be established on the scene of a trench collapse operation:

   a. True
   b. False

UNIT SIX

SOIL PHYSICS

TERMINAL OBJECTIVE

To understand the role that physics and the physical forces associated with soil have in trenching and excavation emergencies.

ENABLING OBJECTIVES

The student will be able to:

- Explain how gravity plays a key role in trench failure

- Describe the term "Unconfined Compressive Strength" as it applies to trenches and excavations

- Define the terms "active" and "passive" soils

- Summarize the effects of water as they apply to soil strength

- Describe how the weight of most soils can be determined mathematically

- Explain how the cubic weight of soil leads to trench failure

- Summarize the most dangerous portion of an un-shored trench, and how a properly shored trench transfers potential energy

If we are going to begin to understand how to handle a trench collapse, a basic understanding of soil physics is necessary. "Stuff don't fall down for nothing." If we strive to understand the concepts related to why a trench collapses, we will go a long way towards preventing one from collapsing further.

One of the most important physical forces of nature that determines whether something stands or falls is **gravity.** Gravity is the force that draws everything to the center of the earth. On a very basic scale, if we dig a hole and leave, it will eventually fill itself back in. This tendency is simply nature's way of reaching the lowest energy state. In reality, a hole in the ground is preventing the earth from

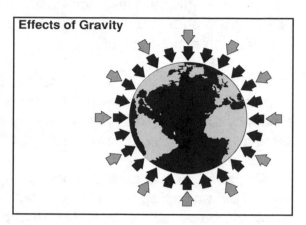

Effects of Gravity

reaching the lowest gravitational energy state, which is an overall spherical planetary shape.

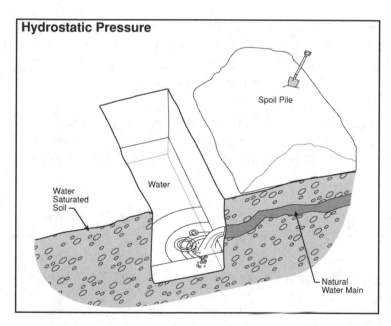

Hydrostatic Pressure

Spoil Pile

Water

Water Saturated Soil

Natural Water Main

Compounding the effects of gravity is the concept of hydrostatic pressure. Hydrostatic pressure can be described as increased pressure caused by the addition of water to the soil profile. As you will find out in a moment, dry soil weighs between 60 and 80 pounds per cubic foot. When the weight of water is added to a very soluble soil, the weight can be phenomenal. In some cases, water saturated soil can weigh as much as 150 pounds per cubic foot.

Going back to the open hole in the ground scenario in the gravity section, imagine the open walls of a trench as pressure outlets. The amount of resistance the soil has to pressure is a measurement of its <u>U</u>nconfined <u>C</u>ompressive

Strength (UCS). In effect, it is unconfined because we have removed the section of earth it was using for stability. When the **unconfined compressive strength** (caused by soil cohesion) is lower than the tension, the soil loses its ability to hold itself up. A higher UCS suggests a more cohesive soil. A lower UCS would indicate a less cohesive or possibly water saturated soil. You will need a good understanding of this concept when we get to the classifications of soils section of the manual.

Another important concept related to the collapse potential of soil is its real or potential tendency to be active or passive. In general, an **active soil** has a tendency to move. This movement may result from the removal or failure of a protective system, or just the inability of the soil to hold its own weight.

A **passive soil** is just the opposite. It has no tendency to move. If you take a yard of dirt and spread it on the ground over a twenty-five-foot radius its potential would be passive at best. Take that same dirt and stand it up as a twenty-five-foot tall column resting on a one-square-foot base and you have generated active soil potential.

Now, let's put this whole physics thing together in a real life scenario. You have just arrived on the scene where a worker has fallen off a ladder into a trench and has broken his leg. The situation is certainly not life threatening, but is aggravating due to his constant bellyaching about the pain.

The spoil pile that is next to the trench is made up of water soaked sand and loam. It appears safe, but you realize that the height of the pile has caused it to have an active tendency. Compounding the problem is the proximity of the spoil pile to the trench lip.

The hydrostatic forces of the water in the spoil, and the spoil pile's proximity to the trench lip are causing a higher than normal compressive force on the trench walls. Additionally, you determine that a very low unconfined compressive strength in the trench walls suggests a water saturated soil that presents a very real collapse potential.

For all these reasons you elect not to enter the trench but instead tell the unfortunate person that he may become a victim of gravity if he does not pick himself up and crawl up the ladder. Using a rope to assist his climb, you pull him from the trench just as a major spoil pile slide occurs.

TIP: Recognition of hazards and procedures for non-entry rescue are an awareness level activity that should be taught to all rescue personnel.

Great job! You just used the concept of soil physics to develop a trench collapse situation risk assessment that may have saved the lives of a number of rescuers and the victim. Man, are you getting good!

PHYSICAL FORCES ASSOCIATED WITH COLLAPSE

The weight of soil and subsequent compressive forces bent this steel beam

Of all the physical factors associated with a trench collapse, probably none is as poorly understood as the weight of soil. For some reason, people just cannot relate to dirt as a volume that has mass and weight. To prove the point, have you ever seen someone stand under a piano being lifted to the upper floor in a building? No! The reason is that if the piano falls, it will crush them to the extent that they could fit in a lady's purse. On the other hand, you do routinely see people standing in a trench beside a volume of potentially collapsible soil, which, by comparison, makes that piano small potatoes!

The weight of 1 cubic foot of most soil is approximately 100 pounds. Yes, that amounts to a 1 x 1 x 1-foot box of dirt. (That figure could be less or more depending on the type of soil and its moisture content.) Certainly more volume than that will fall on you in any significant trench collapse.

To scientifically determine the weight of soil we can make a few observations. Dry soil is one-half soil and one-half air. The specific gravity of rock is 2.65, which means it is 2.65 times heavier than water. If water weighs 62.4 pounds per cubic foot, then rock would weigh 62.4 pounds x 2.65, which equals 165.36 pounds per cubic foot.

What this means is that if the soil you are dealing with is one-half rock and one-half air it weighs 82.68 pounds per cubic foot. If the soil is one-half rock and one-half water (saturated), each cubic foot would weigh 113.88 pounds. We arrived at 113.88 pounds by adding together one-half cubic foot of water at 31.2

pounds, and one-half cubic foot of rock at 82.68 pounds. Generally speaking, dirt weighs between 85 and 150 pounds per cubic foot, which in most situations can be averaged at 100 pounds per cubic foot.

To illustrate the seriousness of the problem, lie down on the floor and have someone sit on your chest. Depending on the size and eating habits of your friend, that would amount to 150 to 200 pounds. Hard to breathe like that, huh? Consider a thousand pounds on your chest.

As discussed earlier under types of collapses, there are specific reasons for how the earth fails in a certain manner. Some of these reasons relate to the manner in which the compressive forces are transmitted down and across a trench.

When the earth is in a stable state, it experiences no unbalanced pressures. That is, at any one point in the earth the pressure is equally distributed. When we cut a hole in the ground the balance is disrupted as pressure is exerted on the remaining soil. When that pressure becomes too great, guess what happens?

So, if normal soil weighs 100 pounds per cubic foot, then a column of dirt 1 x 1 x 6 feet tall would have a total force of 600 pounds per square foot (psf) at the bottom. That means that one column over and one cubic foot up that figure would be 500 psf. This stepping process would continue upwards as 400 psf, 300 psf, 200 psf, and then 100 psf when you got to the top of the column.

The amount of lateral pressure exerted on the un-shored wall is about 33% of the total forces as measured on the bottom of any cubic foot. In a six-foot trench the force at the four-foot level would be approximately 400 pounds per square foot of vertical pressure. The lateral forces that could be expected would be approximately 132 pounds psf. The distribution of lateral pressure occurs on about a 45-degree angle from the bottom of any given plane.

The subsequent release of these lateral forces during a trench wall collapse is evident in the rotational failure. By the way, rota-

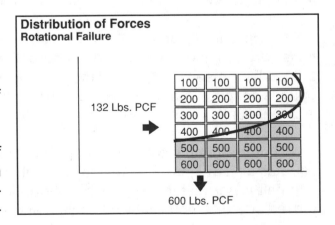

Distribution of Forces
Rotational Failure

132 Lbs. PCF

100	100	100	100
200	200	200	200
300	300	300	300
400	400	400	400
500	500	500	500
600	600	600	600

600 Lbs. PCF

tional failure just so happens to be the most prevalent type of collapse. I hope this is starting to make some sense now.

Actually, the most dangerous portion of the trench wall is the area about one quarter up from the bottom. In a six-foot trench, the highest collapse potential would rest somewhere between the 4 and 5-foot area; that is, if the soil profile is the same throughout the trench wall and no other factor is prevalent. This may be hard to understand until you realize that while the pressure is greater at the bottom, the approximate 90° angle of the bottom at the trench wall provides a measure of stabilization.

The effect of properly stabilizing a trench with shoring takes the pressure from one side of the trench and transmits it to the earth on the other side of the trench. If the pressure is too great for the number and/or sizes of shores in place, they will bend, break, or otherwise collapse, again reiterating Mother Nature's desire to bring stability to the world, one hole at a time.

Summary

In general terms, gravity is the tendency of an object to be attracted to another object. In trench rescue it is gravity, and ultimately its effect on exposed trench walls and or spoil pile, that causes a trench collapse. This collapse can be quite devastating to a worker or rescuer because of the weight of soil. With the addition of water, soil can weigh as much as 150 pcf. For the purpose of estimating soil weight, the rescue industry has established 100 pounds per cubic foot as a standard.

The tendency of a soil to contain energy is called its active or passive potential. Each vertical column of soil creates pressure as measured in pounds per square foot. The deeper the trench the greater the forces as measured at the bottom of any cubic foot of soil.

The wall pressure in a trench can be estimated as 33% of total pressure measured at the bottom of any cubic foot of soil. All factors being the same, the most dangerous portion of a trench is about a third of the way up from the bottom.

The manner in which trenches fail can be attributed to the way the pressure is distributed on the exposed trench walls. The most common type of failure is rotational because of the distribution of force pattern. By properly shoring a trench we transmit the forces from one exposed side of the trench to the other.

Questions (Answers and Discussion on page 243)

1. For the purpose of determining total soil weight we can estimate that a cubic foot of soil weighs approximately:

   a. 25 lbs pcf
   b. 100 lbs pcf
   c. 50 lbs pcf
   d. 75 lbs pcf

2. The most dangerous portion of the unprotected trench wall is the area:

   a. Just below the middle
   b. At the top near the spoil pile
   c. At the bottom of the trench
   d. Just below the top but above

3. The term _____ best describes the effects of mass as it relates to the tendency of one object to be attracted to another object.

   a. Unconfined Compressive Strength
   b. Gravity
   c. Active soil
   d. Angle of repose

4. The amount of resistance the soil has to pressure is a measurement of:

   a. Unconfined Compressive Strength
   b. Passive soil pressure
   c. Active soil strength
   d. Confined compressive strength

5. The effects of water can add a tremendous amount of weight to the excavated material because water weighs approximately:

   a. 52.5 psf
   b. 55 psf
   c. 62. 4 psf
   d. 62.4 pcf

UNIT SEVEN

CONDITIONS AND FACTORS
THAT LEAD TO COLLAPSE

TERMINAL OBJECTIVE

Determine the various factors that can lead to a trench or excavation failure.

ENABLING OBJECTIVES

The student will be able to:

- Explain the effects of water as a factor that can lead to a trench collapse

- Describe the consequences that varying soil profiles and previously disturbed soils can have on open trenches

- List some of the causes of potential vibration that can lead to a trench collapse

- Discuss the spoil pile and its relationship to collapse potential

Many conditions can ultimately lead to a trench collapse. While evaluating these factors, keep in mind that each of them can work in conjunction to generate a very serious collapse situation. None of the factors by themselves will always be the ultimate "straw that breaks the camel's back." Nevertheless, each can, and therein lies part of the problem with a trench rescue. There is no definitive way to determine which one condition, if any, or set of multiple conditions will cause a collapse. Recognizing this fact will be your key to successfully evaluating the factors that could lead to trench failure.

The addition of water can add tremendous weight to a total volume of soil. Water weighs 62.4 lbs. per cubic foot, although the effect it has on the soil is determined by many different factors. For instance, the absorption rate will ultimately determine the total weight for any given volume of soil. Additionally, the effect water has on the ability of the soil to maintain its strength when wet is also critical. Some soils initially gain strength with the introduction of water, then at some point get saturated and become weak. Watch out for soils that look solid but are actually wet and unstable.

Water in the trench is not a favorable factor

Freestanding time of a trench or open excavation can also be a factor that leads to collapse. Once a trench is cut, it is subjected to environmental factors such as drying, wind, and water. The freestanding time also is a ticking bomb with respect to the compressive forces the trench wall can withstand. The longer the trench is open, the closer you are to nature's attempt to fill it back in.

Varying soil profiles

Varying soil profiles within an excavation are a problem rescuers face in determining the classification of the soil and its potential for collapse. Because multiple layers of different materials maintain different friction coefficients, it is often difficult to state with any reasonable certainty that they will react in a certain way. For instance, how much of a fault is created when a layer of sand is sandwiched between two layers of clay? Certainly, the sand creates a slip potential when the earth is not supported on one side of an excavation. Additionally, one needs to determine just how much sand is contained in a specific clay sample and if the sample represents a significant portion of the excavated material.

As in the previous example of sand between layers of clay, water can provide a slip fault in the earth. Frequently, excavated holes contain running water. Because the earth contains these underground streams as a method to move the water from one point to another, they do not just appear, but instead present themselves as problems when we dig the hole. This running water can either be an aquifer, or just the result of the earth releasing the pressure of saturated soil. Both situations present a water removal problem for rescuers involved in this type of collapse scenario.

A plumber working on a catch drain in the late 70s almost lost his life to the hydrostatic forces of saturated soil and running water. In this case, he got one of his legs stuck in the mud at the bottom of a twenty-foot hole. The suction created by the water and wet soil made it impossible for him to pull himself out, and consequently, the more he tried the worse he became trapped. To make matters even worse, water was running in the hole from the pipe he had disconnected at the bottom. Nothing can be as bad as watching a man drown before your very eyes while you are trying desperately to free his submerged legs.

Frantically we assembled all of the dewatering equipment we could get our hands on to try and stay ahead of the impending water. Fortunately for this man, someone had the presence of mind to place an open S.C.B.A. cylinder in the mud between his legs. The introduction of the air offset the suction created by the water just enough for us to pull him out with a rope. Remember, keep dewatering equipment on hand and make sure you have a backup in case the first mud, trash, or centrifugal pump fails.

The water table also dictates what type of rescue situation you will be confronted with at a trench collapse. If you live in an area like Virginia Beach, the mere placing of a shovel in the ground creates a hole for standing water. A high water table means a heavier, more unpredictable soil. Even the most experienced construction workers cannot determine the amount of freestanding time for this type of excavation.

This situation reminds me of the picture of a group of my Phoenix buddies gazing into the bottom of a six-foot Virginia Beach trench that contained two feet of water. "Perplexed" would just begin to describe their reaction to our native soil profile. The bottom line is to know the types of soil in your response area, and preplan your protective system requirements based on this determination.

Previously disturbed soils are another occurrence that is easy to spot, but difficult to interpret. Disturbed soils lack cohesiveness because they are broken and/or mixed with other soil types. You can be certain that if your collapse is in a populated area some type of utility has been placed there before. Rest assured that if you find a cross section of soil in an excavation that contains bricks or bottles, it has been previously disturbed. That is unless we started growing such things and nobody told me!

Heavy equipment location also represents another factor that can lead to a trench collapse. The same equipment that is digging the hole is causing pressure to be exerted on the unsupported trench walls. The rule of thumb in this type of collapse is to turn off the piece of equipment and leave it in place. Generally, this will not be a problem because if it is the piece of equipment that caused the collapse, it will probably be in the hole. If it

Heavy equipment near a trench may increase collapse potential

is not already residing in the hole, leave well enough alone. An important point here is not to let your own heavy equipment become part of the problem. Keep it back outside the rescue area or established hot zone.

There is no hard and fast rule concerning moving heavy equipment once you arrive on scene. Common sense will have to dictate the decision to move equipment or work around it. In all cases you should turn the equipment off once the operation starts and place the keys in a secure location.

Contractors are a funny breed. Their consideration for digging a hole generally revolves around the concept of getting it dug, doing the work, and filling it back in again. It has something to do with making money, although as a lifelong public servant this concept eludes me. However, it does not elude them. The fact is that the contractor can expedite the work process by not moving the spoil pile very far from the trench. The closer the spoil pile is to the trench, the faster it can be used to fill in the trench when the work is complete. This creates a twofold problem for the rescuer. The spoil pile adds additional weight to the unsupported trench, and the spoil pile may slide back into the hole. In any case, if this situation presents itself, the spoil pile will need to be moved far enough back to alleviate the weight concern and provide an area for your ground pads. The moral to this story is to call lots of folks, with lots of shovels!

Another factor that frequently causes a trench collapse is vibration. Vibration can be caused by road traffic near the collapse site or the machinery digging the hole. It can also be caused by augers and other machines used to force utilities under roadways. The key here is to shut the drilling equipment down and limit the traffic not only in the rescue area but also around the general area of the collapse.

It will be necessary for the rescuer to become familiar with all of the factors that can lead to a collapse. Knowing what can cause a potential cave-in will help identify those areas of potential concern present when you arrive on the scene. Ultimately, you will eliminate those concerns that could lead to rescuer injury or death.

Summary

Determining the exact factor that will cause a trench collapse is not possible. What can be evaluated are the factors that collectively contribute to a potential collapse.

Chief among the many factors that may contribute to a collapse is how close the spoil pile is in relation to the trench lip. The extra weight of the soil, particularly if it is a water saturated spoil pile placed too close to the trench, can be too much for the unconfined walls to hold. An additional problem in trying to evaluate this situation is that some water saturated soil looks strong and cohesive.

Another factor that needs close attention is the amount of time a trench has been open. The freestanding time of a trench allows the wind and other elements to weaken the exposed trench walls. Other factors that may need to be considered are varying soil profiles, previously disturbed soils, heavy equipment location, and vibration.

Questions (Answers and Discussion on page 244)

1. Which of the following can be considered a factor that can contribute to a trench collapse:

    a. Freestanding time
    b. Previously disturbed soils
    c. Hydrostatic forces
    d. Varying soil profiles
    e. All of the above

2. A good indication that soil has been previously disturbed is:

    a. Varying soil layers are evident
    b. Foreign material is in the spoil pile
    c. Utilities are already located in the trench
    d. B is correct
    e. B and C are correct

3. You should always move heavy equipment from near the trench:

    a. True
    b. False

4. Vibration is a factor that can cause a trench collapse and can be attributed to:

    a. Drills that make utility holes under roads
    b. Nearby traffic
    c. Construction equipment located on the scene
    d. All of the above

5. Of the factors that can contribute to a trench collapse, the amount of time the trench is open is a major factor. The time the trench is open is:

    a. Open time
    b. Free open time
    c. Wall free time before collapse
    d. Freestanding time

UNIT EIGHT

TYPES OF TRENCH COLLAPSES

TERMINAL OBJECTIVE

Describe the various types of trench collapses and why they occur.

ENABLING OBJECTIVES

The student will be able to:

- Describe the differences between a spoil pile slide, slough failure, shear wall collapse, toe failure, wedge failure, and rotational failure

Collapses can be somewhat predictable based on soil profile, the type and size of the trench, and conditions under which the trench was excavated. Familiarity with the types of collapse will help you determine the trench's potential for collapse, and the proper protective system appropriate for making it safe.

The spoil pile slide is the result of excavated earth placed too close to the lip of the trench. This type of collapse is not as common as you might think, as most contractors recognize this hazard.

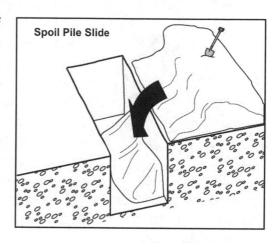

A spoil pile slide occurs when the soil's natural **"angle of repose"** becomes greater than its cohesive tendency. When this happens, the spoil pile slides back into the opening. Another factor creating this situation is that newly excavated dirt may have a certain amount of moisture content that provides some cohesiveness. As the soil dries it becomes less stable. Remember, as we suggested earlier, a hole in the ground wants to naturally fill itself back in.

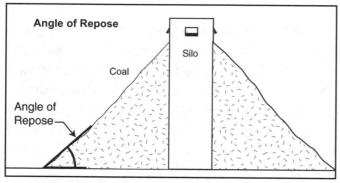

This concept demonstrates the angle at which a substance will settle under its own weight

A slough failure is the loss of part of the trench wall, and it can be the result of several conditions. Frequently, the force associated with unconfined hydrostatic pressure becomes greater than the soil's ability to stand. It can also be caused by a spoil pile being placed too close to the trench lip. As extra earth is piled up, its weight is transmitted in a downward force communicated through the trench walls. When this pressure exceeds the soil's ability to support it, a failure will occur. Cracks in and around the excavated

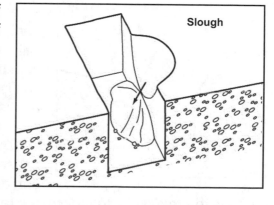

surface and multiple soil layers are key indicators that you may have the potential for a slough collapse.

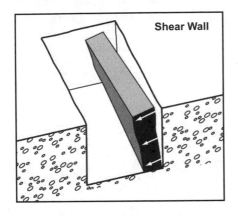

A shear wall collapse occurs when a section of soil loses its ability to stand and collapses into the trench along a mostly vertical plane. This condition can be caused by cracks in the earth's surface exposed to the weather over time. As the water runs into the opening, it washes out dirt and then dries. Over time, this washing and drying action causes the hole to become deeper and deeper, until it is not supported on two sides and a wall of dirt falls into the trench. Shear wall collapses are normally associated with fairly cohesive soils. That factor makes them look safe. Big problem! As you might imagine, this type of failure can create a big time collapse situation.

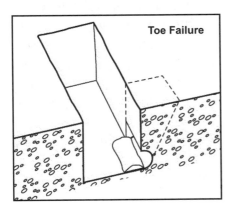

A Toe failure is a slough that occurs at the bottom of the trench where the floor meets the wall. As the soil falls into the trench it creates an opening at the bottom that is characteristic of a cantilever. It can be caused by a sand pocket, or the effects of water at the bottom of the trench.

This is a very dangerous type of failure for many different reasons. First, the rescuers might not notice the toe failure until they are standing on top of the cantilevered section of earth. It then becomes entirely possible that they will get an "up close and personal" view of the hole by virtue of being in it. Secondly, the situation is hard to correct until after a protective system is in place.

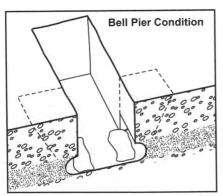

The effects of water accumulation can also cause a **bell pier condition**. This type of situation does not usually occur suddenly, but more often is the result of a long-term toe failure on both sides of the trench floor. The bell pier condition is also dangerous for the reasons previously discussed.

A rotational failure is characterized by a scoop shaped collapse that starts back from the trench lip and transmits itself to the trench wall in a half moon shape. These types of failures can result in the movement of large sections of soil to the trench floor. What remains is a collapse that looks like someone took a spoon and carved out a chunk of earth. If the rotational failure is large enough, it creates a difficult protective system problem for rescuers, since the void will need to be filled at some point in the operation.

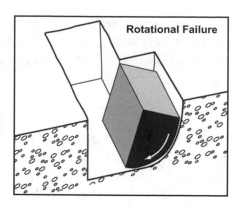

Rotational Failure

The last type of collapse we will discuss is the **wedge failure**. This type of failure normally occurs with intersecting trenches. It is characterized by an angle section of earth falling from the corner of two intersecting trenches. The wedge failure can be sudden and catastrophic. You will learn how to manage intersecting trenches later in this manual.

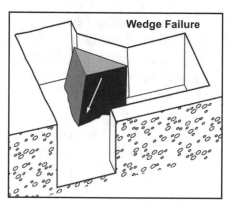

Wedge Failure

While it may be relatively simple to determine what type of collapse has occurred, it is often quite difficult to understand why the collapse has taken place. Commit each of these types of collapse to memory; then as we discuss the physical forces associated with various collapse scenarios, the reason each happens will become clearer.

Summary

While the conditions that lead to collapse might be somewhat unpredictable, the manner in which trenches collapse is fairly predictable. There are several types of collapses that normally occur given a set of conditions and based on the type of excavated opening.

The Spoil Pile Slide occurs when the angle of the soil is greater than the soil's cohesive tendency. The soil would then fall into the trench until the pile reached its angle of repose or the angle at which a product will stand under its own weight and not slide.

The Slough failure is a failure of part of the trench wall from under the top and subsequently can produce a cantilever situation of the lip. A Shear Wall collapse usually happens along a vertical plane. This type of collapse is usually caused by cracks in the earth that are exacerbated over time due to environmental factors.

The Toe Failure can be characterized as a slough that occurs at the toe of the trench. When this situation presents itself on both sides of the trench wall at the toe it is called a Bell Pier Condition. Sandy, water saturated soil frequently causes this type of failure.

The Rotational Failure is a scoop shaped collapse that starts back from the trench lip and ends up as a half moon shape. Frequently the pressure of the spoil pile or some other additional weight to the trench lip is the cause of this type of failure. Additionally, it could be caused by varying soil profiles contained in the trench wall.

The Wedge Failure generally occurs in intersecting trenches at the point where the trench intersects. It is characterized by an angle section of earth falling from the corner of an intersecting trench.

Questions (Answers and Discussion on page 245)

1. When the excavated material falls back into the trench it is best described by the term:

   a. Shear wall collapse
   b. Spoil pile slide
   c. Slough failure
   d. Rotational failure

2. A collapse in which the material loses its ability to stand and fails along a mostly vertical plane best describes a:

   a. Shear wall collapse
   b. Spoil pile slide
   c. Slough failure
   d. Rotational failure

3. The most common type of soil failure in an intersecting trench is:

   a. Wedge
   b. Slough
   c. Shear
   d. Drop

4. A condition that presents itself as a failure at the toe and on both sides of a trench is:

   a. Toe failure
   b. Slough failure
   c. Cantilever collapse
   d. Bell pier condition

UNIT NINE

SOIL CLASSIFICATION AND TESTING

TERMINAL OBJECTIVE

To understand how consistent standards are used to classify soil
based on its potential for collapse and the methods used
to test soil for classification purposes.

ENABLING OBJECTIVES

The student will be able to:

- Describe the four classifications of soil

- Explain the parameters that lead to individual classifications

- Describe the various methods used to perform visual and manual soil testing

- Explain the proper use of penetrometer, shearvane, and torvane soil testing instruments

SOIL CLASSIFICATION AND TESTING

The system of classifying soils is a hierarchical approach to determine the performance of a soil based on a decreasing order of stability. In a nutshell, some general assumptions are made about what products are in the soil and how they can be expected to behave when excavated.

Each type of soil represents a varying degree of danger based on the characteristics that make it a part of that class. In multiple layers, the classification will be determined by the layer that is normally least stable.

All classifications need to be determined based on one visual and one manual test performed by a competent person. Being able to classify the various soil types will allow the rescuer to pick the appropriate protective system, and more importantly, gauge the risk involved in any potential collapse situation.

STABLE ROCK

The least dangerous, from a collapse perspective, is stable rock. This type of soil is a natural solid material that can remain standing after excavation. The danger associated with stable rock excavations generally is from anything but a collapse. This does not mean that some excavated products cannot fall on a worker. As such, accidents in this environment are usually from worker falls, or equipment failures that cause entrapments.

TYPE A

Type A soils are cohesive materials with an unconfined compressive strength of 1.5 tons per square foot (tsf) or greater. Examples of this type of soil include: clay, silty clay, clay loam, and sandy clay loam. Cemented soils are also considered Type A. However, no soil is Type A if:

1. The soil is fissured
2. The soil is subject to vibration
3. The soil has been previously disturbed
4. The soil is part of a sloped soil layer that is greater than 4 horizontal to 1 vertical or
5. The material is subject to other factors that would require it to be classified as a less stable material

TYPE B

Type B soils are those cohesive materials with an unconfined compressive strength greater than 0.5 tsf but less than 1.5 tsf, or a granular cohesion-less material including angular gravel, silt, silt loam, sandy loam, and sandy clay loam. Type B may also be a previously disturbed soil, unless it would otherwise be classified as Type C. It may also be a soil while meeting the unconfined compressive force requirements for Type A, is fissured or subject to vibration. In addition, it could be a material that is part of a sloped system steeper than 4 horizontal to 1 vertical.

TYPE C

Type C soils are those cohesive materials with an unconfined compressive strength of 0.5 tsf or less. This includes granular soils, sand, and sandy loam. Type C soils are also submerged soil, or soils from which water is freely flowing, or submerged rock that is not stable. Additionally, this includes sloped or layered systems where the layers dip into the excavation at a slope of 4 horizontal to 1 vertical or steeper.

It should be noted that once a soil has been classified and conditions that determined the original classification change, a reclassification must be done by a competent person. This may necessitate a change in the type of protective system selected to accomplish the rescue, or at least a change in the risk/benefit "evaluation."

C- 60 Soil

There is a good possibility that during your education in trench rescue you will come upon another classification of soil that is not listed in the OSHA Appendix. C-60 Soil Classification was developed by Speed Shore to identify soil that is a "moist cohesive soil, or a moist dense granular soil which does not fit into Type A or Type B classifications, and is not flowing or submerged. This material can be cut with near vertical sidewalls and will stand unsupported long enough to allow shoring to be properly installed."

This is allowed because OSHA recognizes the use of classification tables other than those provided in the OSHA standard. It is only permitted if the tabulated data is approved by a registered professional engineer for use in design and construction of the protective system. The key here is the term "registered professional engineer." Based on this OSHA interpretation, you may indeed hear of additional sub categories of recognized soil types.

SOIL TESTING PROCEDURES

VISUAL TEST

The visual testing requirements to effectively determine a soil classification are accomplished by inspecting the excavated material, the soil that forms the trench wall, and the excavation site in general. This overall site evaluation will help offset the fact that the layers in the dirt may change as the excavation is dug deeper. Remember that soil classification is based on the weakest soil in a layered system.

When the dirt is being excavated, or when you are arriving on the scene, observe the soil that has been previously excavated. This will help you determine the initial cohesiveness of the soil. Soil that is made up primarily of fine-grained material that remains in clumps is said to be cohesive. Soil that breaks up easily and is primarily composed of coarse grained sand or gravel is granular.

The trench soil particles will tell you a lot about the soil composition, but the most important area of the visual assessment will be the trench walls and the area surrounding the trench lip. On the trench walls, look for layered soil and any indication that the soil was previously disturbed. The presence of utilities can indicate disturbed soil. As previously mentioned, a mixed soil will usually not be cohesive. In general, "similar" or "identical" particles of soil are the most likely to be attracted to each other and to remain attracted.

A fair visual evaluation also considers if the trench wall contains fissures or tension cracks that could suggest a potential collapse. Openings, or spalling, in the exposed trench are indicators that the walls are under tension and subject to rapid release and subsequent collapse. The area around the trench should also be checked for cracks in the soil. This would indicate possible soil movement caused by the trench walls falling into the trench, which in turn creates voids in the earth surrounding the walls.

The hydrostatic forces can also be analyzed by looking for indications of standing, seeping, or running water. Water adds weight and weight adds more tension to the trench walls. This is especially true of surface water that has pooled near the trench opening. As a clue to the anticipated hydrostatic forces, look for indications that the contractor has "well pointed" the area surrounding the excavation. Well points are used to remove excess water from saturated soil before digging a trench.

MANUAL TEST

A manual test is necessary to determine the various characteristics of the soil, and to learn its relative strength when placed under force. These types of tests are used to formulate an assumption regarding the material's overall ability to free stand.

PLASTICITY TEST

The plasticity of the soil is the property that allows the soil to be deformed or molded without appreciable change in total volume. The test is done by molding a moist or wet sample into a ball and then attempting to roll it into threads as thin as 1/8-inch in diameter. A cohesive material can be rolled into threads without crumbling. As a rule, if a two-inch length of 1/8-inch thread can be held on one end without it tearing, the soil is considered to be cohesive.

RIBBON TEST

The ribbon test is used to determine how much clay or silt the soil contains. This test is done with a saturated, fine soil and fine sands that are rolled together between the palms of the hands until a cylinder approximately 3/4 inch thick by 6 inches long is formed. The cylinder is then placed across the palm of the hand and squeezed between the thumb and forefinger until it is approximately 1/8 inch thick. The squeezed portion is then allowed to hang over the side of the hand. If the cylinder forms six (or more) ribbon sections in length or longer it is said to be clay. The longer the ribbon, the more clay the soil contains. If it forms shorter broken ribbons, then the soil contains silt. A clay loam will barely form a ribbon of any length.

DRY STRENGTH TEST

A dry strength test is done to determine the propensity of the soil to fissure. If the soil is dry and crumbles on its own (or with moderate pressure) into individual grains or fine powder, it is granular. If the soil is dry and falls into clumps that in turn break into smaller clumps that can only be broken with difficulty, it may be clay in any combination with gravel, sand, or silt. If the dry soil breaks into clumps that do not subsequently break into smaller clumps, initial clumps can only be broken with difficulty, and there is no visual indication the soil is fissured, the soil may be considered unfissured.

THUMB PENETRATION TEST

The thumb penetration test can be used to estimate the unconfined compressive strength of cohesive soils. This type of test is accomplished by placing your extended thumb against the exposed material and attempting to push through the sample. Type A soils can be readily indented with the thumb, but only penetrated with great effort. Type C soils can be easily penetrated several inches by the thumb and molded with little effort. Please note that this test should be completed as soon as practical after excavation occurs in order to keep the drying influences of the environment from affecting the sample.

DRYING TEST

The drying test is used to determine the difference between cohesive material with fissures, unfissured cohesive material, and granular material. The procedure involves drying a sample of soil that is approximately six inches in diameter and one inch thick until it is thoroughly dry. If the sample develops cracks as it dries, significant fissures are indicated. Samples that dry without cracking should be broken by hand. If considerable force is required to break a sample, the soil has significant cohesive material content. The soil should then be classified as an unfissured cohesive material and the unconfined compressive strength should be determined. If a sample breaks easily by hand, it is either fissured cohesive material or a granular material. To distinguish between the two, pulverize the dried clumps of the sample by hand or by stepping on them. If the clumps do not pulverize easily, the material is cohesive with fissures. If they pulverize easily into very small fragments, the material is granular.

PENETROMETER AND SHEARVANE

There are a couple of instruments that can be used to determine the unconfined compressive strength of the soil sample. The most popular of these devices is the pocket penetrometer and shearvane. When used correctly, the amount of force required to insert the instrument into the trench wall correlates to an unconfined compressive strength of that section of sample. The soil must have some moisture content to extract an accurate reading, primarily because the instrument must be pushed into the wall or sample and read numerically.

Pocket penetrometer

All of a soil's characteristics require proper evaluation to assume any degree of confidence in your protective system, and thus the safety of your rescuers. To illustrate how critical testing and proper evaluation can be, look at trench fatality statistics and subsequent soil profiles. An overwhelming majority of trench fatalities occur in clay and mud excavations. This is because these soils appear safe as compared to sand or gravel. Remember, do not draw conclusions about a trench's safety based on its appearance. Rely on your newfound knowledge of testing procedures.

Summary

Soils are classified and listed in decreasing order of stability as Stable Rock, Class A, Class B, or Class C. Each class is determined by the soil's content, characteristics displayed, or unconfined compressive strength as determined by testing.

Either manual or visual testing is used to determine soil classification. Visual testing is done by inspecting the excavated material for its predominant composition and moisture content. In addition, the area around the trench and the trench itself are examined for characteristics such as clumping or cracking of soil. Manual testing is done by performing a Plasticity Test, Ribbon Test, Thumb Penetration Test, Drying Test, or by using a mechanical device like a penetrometer, shearvane, or torvane-testing device.

Questions (Answers and Discussion on page 246)

1. When classifying soils, which of the following is not an approved testing method?

    a. Visual
    b. Manual
    c. Plasticity
    d. Water drip test

2. A test that is performed by rolling the soil threads, which determines the soil's ability to be deformed and molded best describes:

    a. Ribbon test
    b. Thumb penetration test
    c. Manual test
    d. Plasticity test
    e. None of the above

3. Warning signs of potential collapse consist of:

    a. Fissures
    b. Drying of exposed walls
    c. Falling or loose debris falling from the walls
    d. All of the above

4. You are performing a quick evaluation of the soil at a collapse operation and notice that the excavated material is dry, cracked, and granular. As a manual test you attempt to roll the soil in your hand which causes it to break and crumble. Based on your assessment this would be best classified as:

    a. Type A
    b. Type B
    c. Type C
    d. Stable rock

5. The trench you are examining is ten-feet deep and four-feet wide with soil that is a clay loam mixed with sand. The upper portions of the trench have water seeping from the walls with an additional two inches of water in the

bottom of the trench. Based on your assessment you conclude that this type of soil is:

a. Type A
b. Type B
c. Type C
d. Stable rock

6. Cohesive materials that have an unconfined compressive strength greater than .5 tsf but less than 1.5 tsf best describes:

a. Type A
b. Stable rock
c. Type C
d. Type B

7. The test that is used to determine a soil's propensity to fissure is:

a.   Ribbon test
b.   Thumb penetration test
c.   Manual test
d.   Plasticity test
e.   None of the above

UNIT TEN

PERSONAL PROTECTIVE EQUIPMENT
FOR TRENCH RESCUE OPERATIONS

TERMINAL OBJECTIVE

To describe the proper personal protective equipment
used in trench rescue operations.

ENABLING OBJECTIVES

The student will be able to:

- Discuss the advantages and disadvantages of fire fighting "turnout" gear, jumpsuits, and regular long sleeve pants and shirts for trench rescue

- Explain the pros and cons of the various types of hand protection

- Describe the preferred helmet for trench rescue operations

- Determine the correct type and use of eye protection

- Recite the preferred level of foot and ankle protection for the trench environment

- Describe the specialty equipment that may be required during a trench rescue

- Summarize the benefits of developing a team culture that maintains safety as a top priority

In every case, the safety of your personnel is dependent on their level of skill, as previously developed during training and real time incidents. This training and experience may give them an initial advantage in the rescue effort. However, if they do not have the proper PPE and are subsequently injured in the rescue attempt, overall the operation will not be considered a success. In other words, all of the training and experience in the world is hardly worth its salt if personnel do not have the proper personal protective clothing and equipment to give them reasonable protection during the rescue.

We can divide personal protective equipment into two main categories—the clothing that rescuers wear to protect themselves and the equipment provided that would not be considered standard issue items. This would correlate with the difference between a pair of gloves and a breathing apparatus.

This safety part of the rescue effort is not rocket science, but more like common sense. For the sake of argument and so that all of us are on the same page, let us review appropriate trench rescue clothing and equipment.

STANDARD ISSUE EQUIPMENT

The rescue personnel on the scene of a trench collapse need a variety of protective clothing to minimize the effects of the weather and any related trauma caused by working in, and around, machinery and tools. At a minimum, rescuers need a jumpsuit or other long sleeve shirt and pant combination, gloves, steel-toed boots, helmet, and eye protection. Other items, depending on the environment expected, could include hearing protection and safety vests for visibility concerns.

CLOTHING

In all cases, rescuers need to have some level of protection from skin abrasions due to contact with on-scene objects. If it is very cold outside this might be a set of fire suppression turnout gear. While **fire-fighting gear** is very bulky, and not recommended as standard protection, when someone is cold and miserable, concentrating on what you are doing is difficult. The truth is that turnout gear is appropriate in some circumstances, and if that is all you have, make sure you have it on. On the other hand, in colder weather, it may be a good call to have your personnel layer in many thinner types of outer clothing. The key is to evaluate the advantages of the warmth the gear provides, as compared to the disadvantages of not having as much manual dexterity.

The **standard jumpsuit** is more than adequate skin protection for most trench emergencies. It will cover the arms and legs and can be ordered with the name of your organization, and reflective tape for visibility purposes. The disadvantage of the jumpsuit is that it retains heat because of its one-piece design, and something else that I call "crotch creep." If the jumpsuit is not large enough, when you reach over your head, suddenly you're "talking soprano." If it is too large, the crotch hangs down between your knees and creates a walking inconvenience. Remember, we want the rescuers to be as comfortable as possible so they concentrate on the efforts at hand and not their clothing. This alone makes a case for providing each member of the rescue team with their own standard issue of protective clothing.

In any case, the minimum level of protective clothing should be **long pants and long sleeve shirts**. You will notice there was no mention of short pants or short sleeve shirts. If it is hot, rehab often and set up misting and other ventilation fans to cool the area. If the outside temperature is that much of a consideration, set a tarp over your trench to shield the rescue area from the sun. Whatever you do, make sure you wear the proper protection for the environment in which you are working.

GLOVES

The glove is the undoing of the rescuer. We know they have to be worn, but performing any rescue-related function is difficult when wearing them. What normally happens is that when rescue personnel are standing around talking they have their gloves on. The first time they are required to do something, off come the gloves. Frequently, I remind rescue personnel that gloves will afford their butts only limited protection when stored in their back pockets.

Here again, I think that comfort is the key to success when encouraging your personnel to wear their gloves. Fire fighting gloves are designed for fighting fire and, therefore, are not comfortable when a person is trying to utilize tools or operate equipment. On the other hand (no pun intended), if all they are going to do is move lumber, the fire-fighting glove may be appropriate.

The truth is that the best kind of glove for trench rescue personnel is the standard run-of-the-mill leather garden glove. This type of glove is sturdy enough to keep abrasions and splinters to a minimum, while still comfortable enough that rescuers will generally keep them on their hands while working.

If you have plenty of money, or there is a military installation in your area, you may want to try a pair of nomex flight gloves. This glove is very pliable and has a thin layer of leather in the palm. It is, by far, the most comfortable of all gloves to wear, and you can still accomplish just about any function with them on. On the down side you may want to do some pricing before committing to this type of glove. They are expensive when compared to the regular leather glove. On the upside, they can be used for other specialized rescue functions like confined space.

HEAD PROTECTION

The most critical piece of protective clothing that you will wear during a trench rescue is the **helmet**. Unlike the glove issue, we are not talking about the occurrence of a splinter or abrasion if you do not wear it. What we are talking about is head trauma from a hammer, shore, or piece of lumber—something from which you do not easily rebound and which could in fact prove to be a career-ending mistake.

As with fire fighting clothing, most rescue/fire personnel will have a fire helmet. I am almost positive that when you request special gear for your rescue team, some bean counter will balk at buying team members special helmets when others are already provided. The problem with fire helmets is the same as with fire gloves. If it is heavy and uncomfortable, the rescuers will take it off. This is not what we want. Tell the bean counter that we are not fighting fires in the trench, and if one has the potential to break out, we will be in the proper gear.

Taking into account what we are trying to protect ourselves from, the best helmet to wear at a trench emergency is the heavy-duty construction helmet. It will provide limited impact protection, is lightweight, and for the most part is balanced. There are many varieties of this type of helmet available, both commercially for general industry, and through rescue gear manufacturers. Just make sure the helmet is ANSI approved for the task and has a chinstrap to keep it from falling off when the wearer bends over.

TIP: Frequently you will need to remind people to wear their helmets while on the rescue scene. A technique I have found effective in reminding them to wear their helmets when they have found occasion not to, is to tell them that you know a great place to store their helmets. When they ask where, you tell them: "on their heads." A little embarrassment along with some humor goes a long way in a tense situation.

EYE PROTECTION

Eye protection is essential for trench rescue work. Something needs to provide a barrier from flying nails and other debris that can be expected in a construction environment. The only rule for eye protection is that it needs to encompass the entire front of the eye, and not change positions when the head is moved. Therefore, helmet-mounted eye protection is not recommended.

Full face goggles, while providing the most protection, will usually fog up when the rescuer gets hot. When that happens, what do you think comes next? That's right...the goggles end up hanging around the neck and not over the eyes—again providing you an excellent opportunity to tell someone of the proper place they can be stored.

The **standard pair of safety glasses** is more than satisfactory for trench rescue. Most can be purchased with or without the sun or glare protection, and they will stay in position when there is movement of the rescuer's head. In addition, they will not usually fog up because ventilation is allowed around the outside of the lenses. As with the helmet, make sure any safety glasses you use are approved for trench rescue purposes.

FOOT PROTECTION

This area of protective clothing is a no brainer. All personnel should be wearing **steel-toed, steel-shanked boots**. Not only is there every opportunity to drop something on your foot, but nails and other sharp debris are typically found everywhere. In addition, it is also a good idea to have a high top boot. This will provide a measure of ankle support in the event you step on a piece of lumber or other equipment and turn your ankle.

SPECIALTY ITEMS

Respiratory protection should always be considered at the scene of a trench collapse. At the very least, rescue personnel will be subjected to flying dust and dirt while working in and around the trench. As a minimum you may want to consider the use of a dust mask. In any case, if there is any indication that an atmospheric problem might exist, you will want to have SCBA or SABA before entry into the environment.

Hearing protection is a good idea for anyone working around compressors or saws. Choose a level of protection that will shield the rescuer's ears from high noise frequencies, but will not block out all communication. It is more dangerous if the rescuer cannot hear anything than for them to be exposed to loud noises.

Welders wear **skullcaps** to keep their head cool under their helmets while welding. Rescue personnel have started wearing them recently for the same purpose. As the heat from your head causes perspiration, it wets the skullcap, which in turn keeps the head cooler. You can also wet the skullcap before you put it on to get a head start on the process.

Leather chaps are good leg protection and a great idea if anyone is going to be doing much cutting with a chain saw. While they will not keep someone from sawing their leg off, chaps might be enough to stop or deflect a chain that bounces off the wood and inadvertently hits a rescuer's lower body.

DEVELOPMENT OF A SAFE CULTURE

Instilling a safety mindset into the culture of your team is not something that happens overnight. It is the result of many hours of training and the discipline that comes from everyone on the team being accountable. It occurs because individual team members demand it from each other.

When an injury brings down a team member, everyone suffers the loss. There is nothing wrong with being safe, or demanding that the team maintains a positive attitude about being safe. Ultimately it is something to be proud of and it should be emulated by other teams that deliver specialized rescue services to the public.

Summary

Trench rescue is essentially construction work with a little technical rescue thrown in for good measure. In any case, fire fighting turnout gear is generally not appropriate unless weather conditions call for its use. A jumpsuit or long sleeve shirt and pants are the best protection for the trench rescue environment.

Hand protection should always be a compromise between rescuer safety and comfort. The general all-purpose leather glove, while not providing a high level of protection, is generally comfortable enough that rescuers will keep them on while working. The same can be said of helmets. Find a helmet that is light, balanced, and will protect someone's head from a falling object or direct bump but is also light enough that they don't take it off while performing work.

Eye protection should protect the eyes from flying debris and nails. While goggles will work for this purpose, they often become fogged up from the rescuer's body heat and then end up around their neck. Regular ANSI approved safety glasses will provide enough protection from most hazards.

Never compromise on footwear. Ankle height steel-toed and shank boots are a must and will protect from most penetrating objects. In addition, they will provide a measure of ankle support.

While evaluating the hazards presented by a trench rescue, there will almost always be a case for using specialty equipment like breathing apparatus and hearing protection. Use the tools that will protect your folks best, as they are your number one investment. In all cases, safety on the scene of any technical rescue incident is a culture that is imparted by team leadership and engrained into the fabric of your rescue team.

Questions (Answers and Discussion on page 248)

1. The minimum level of personal protective equipment required for trench rescue operations is all of the following except:

   a. Eye protection
   b. Gloves
   c. Steel toed boots
   d. Skullcap
   c. Helmet

2. Fire fighting gear should never be worn on a trench collapse.

   a. True
   b. False

3. The minimal level of personal protective equipment worn by rescuers at a trench incident is:

   a. Helmet, gloves, and boots.
   b. Determined by the specific hazards presented.
   c. Helmet, gloves, boots, and eye protection.
   d. None of the above

4. Eye protection should:

   a. Be ANSI approved
   b. Not interfere with line of sight
   c. Be comfortable for rescuers to wear
   d. Protect from most flying debris
   e. All of the above

UNIT ELEVEN

EQUIPMENT AND TOOLS FOR TRENCH RESCUE OPERATIONS

TERMINAL OBJECTIVE

To describe the types of equipment and tools used in trench rescue operations.

ENABLING OBJECTIVES

The student will be able to:

- Explain the use of ground pads for trench rescue

- Describe how sheeting is used in trench protective systems

- Identify the various types of shores used in trench rescue and how each works

- Describe the various types of tools used in trench rescue operations

- Explain the use of various trench rescue tools utilized in collapse operations

There are many different types of tools and equipment that may be required on the trench collapse scene. Most individuals are already familiar with much of this equipment, so the majority of this section will be dedicated to the utilization of less familiar items.

GROUND PADS

The area around the trench lip is a very unstable area. Any additional weight on the soil could cause a secondary collapse of the trench. For this reason, **ground pads** are used to line the area around the trench after the removal of the excess dirt from the spoil pile. The effect of the ground pads is to distribute the rescuer's weight over a larger area. To get a greater understanding of the pressure per square inch concept, have a woman stand on your foot. Now, have her put on a high-heeled shoe and do the same thing. Ouch! That is concentrated pressure. The same principle is involved when you stand in one place on the lip of a trench without the benefit of a ground pad to spread out your weight.

There are several types of ground pads that can be used to accomplish the same purpose. The most common type of ground pad is a 4 x 8 sheet of 1/2-inch plywood. This type of ground pad provides a large area to distribute weight, and provides a fairly nice platform from which to work. The drawback to this type of ground pad is that it covers a significant portion of the ground around the trench, which makes it diffi-

4x8 ground pads

cult to inspect for deteriorating conditions. In addition, if you have a large spoil pile it can take a long time to move enough dirt to get the ground pad flat.

TIP: When covering up cracks on the trench lip with ground pads, mark the area on top of the ground pad with paint. This will remind folks not to stand on that area and you can also check for worsening conditions.

Marking ground pads to denote cracks in the soil under the pad

In some cases a 2 x 6-inch piece of lumber, usually 10 or 12 feet long, can be used as a ground pad. Actually, any 2-inch material can be used. The 2-inch material does the same thing as a piece of ply-wood, only the weight is distributed over a smaller area. A disadvantage of this type of ground pad is that it is small and hard to maintain a stance on. Fre-quently, you get the feeling of tippy-toeing around the trench. That in itself can create a trip and fall hazard. The main advantage of the 2 x 6-inch ground pad is that less spoil pile has to be moved to facilitate its use.

As you can see, each type of ground pad has good and bad points. Whichever ground pad you decide to use, place it in service by staying on the safe side of the pad and not between it and the trench. Clear a small area, then push it in front of you until you can get to another area. Always stay on the ground pad.

SHEETING

Sheeting material can be interconnected steel uprights, sheets of plywood/timber, or ShorForm panels that are used to contact the walls of the trench. They function as a shield system with uprights by holding back running soil and debris. In hard pack soil, sheeting is not required as a part of the protective system. However, in loose or shifting soil it is necessary and preferred for hold-ing back running material.

Most of what you will come in contact with is ShorForm, FinnForm or **home-made sheeting** panel. The **ShorForm panel** offers the rescuer a viable and safe sheeting panel for efficiently shoring trenches. It is a high strength, rela-tively lightweight, and non-conductive material that is made entirely of arctic white birch. The exterior of the ShorForm panel is made durable by phenolic resins, which are impregnated into the hardwood surface to provide for maxi-mum re-use and ease of cleaning.

It should be stated here that it is perfectly acceptable to make your own panels con-sisting of two pieces of 4' x 8' 3/4-inch exterior grade ply-wood. By gluing the panels together you create a hefty 1-1/2-inch piece of plywood that, when assembled with a strongback, is quite sturdy but a chore to carry and place.

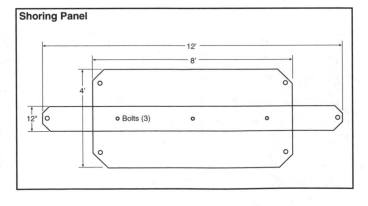

Shoring Panel

Whatever type of sheeting you choose, there are a couple of things you can do to facilitate ease of use and storage. Cutting all of the corners off at ninety degrees reduces the possibility of splintering should the panel be dropped on the corner or edge. Hand holes or holes for ropes can also be drilled to aid with the placement and adjustment of the panels in the trench.

All panels that are placed have to be used with a 2" x 12" x 12-foot upright commonly referred to as a **strongback**. The strongbacks are bolted to the panels using $3/4$" x $3\text{-}1/2$" machine bolts and they are the main component in the protective system. The strongback is the part of the protective system that transmits the forces across a vertical plane into the trench wall. Care in assuring that the panels are set tight

ShorForm Panel

against the walls of the trench will assure that the strongback can transfer the necessary force from the shore to the trench wall.

TIP: Take a section of rope and place the end through a hole in the strongback side of the panel on the bottom. Tie a knot in it so it won't pull back through, then do the same for the other side. By lowering the panel into the trench, while pulling back on the ropes, the panel will get a very nice vertical set against the trench wall.

Using a rope around the strongback to adjust panel

SHORES

Shores are the component in the protective system that carry the force from one side of the trench to the other. They are represented by a number of different materials and forms, each having its own limitations.

The most common (and oldest) type of shore is the **timber shore**. Timber shores usually are 4" x 4", 4" x 6", and 6" x 6" sections of Douglas Fir with a

Fire fighter placing a timber shore

bending strength of not less than 1500 psi. Timber can be used in all trenches that do not exceed 20 feet in width in order to maintain a relatively high strength to length ratio which gets stronger the shorter the shore is cut and weaker the longer the shore is when placed. Timber shores have the advantage of being low cost when compared to other shores on the market. In addition, they can be cut to varying lengths.

The size and length of the timber shore selected is based on the depth and width of the trench as determined by the type of soil present. Appendix C of the OSHA standard contains information that can be used when timber is the type of shoring material selected for use in the protective system. A word of caution is necessary at this point if you are considering timber shoring for a protective system. If you are in any trench that is greater than 10 feet in depth, and over 4 feet in width with any kind of soil, the minimum timber shore size is 4" x 6" and in the case of Type C soil it is 8" x 8". That, my friends, is a big piece of wood that most lumber yards have to special order. The moral to the story is, don't take for granted that an 8" x 8" timber of suitable length will be at the lumberyard when you have a trench collapse.

Pipe Screw Jacks

Ellis Screw Jack

Screw jacks are a common tool used in conjunction with timber shoring to form a tight wall to wall shore. This type of shore has a boot end, which fits over a piece of wood and then can be tightened by a thread and yoke assembly. They are also sometimes referred to as pipe jacks when used in conjunction with varying lengths of pipe. Screw jacks are relatively inexpensive, although not very strong when compared to other types of shoring. Particular care must be taken not to over extend this type of shore.

Hydraulic shores, like those manufactured by Speed Shore, offer a type of protective system that contains the shore and the upright as one unit. The system is lowered in the trench from the top, and then expanded by using a 5-

Hydraulic Speed Shore

gallon reservoir of non-flammable and bio-degradable fluid. After the shore is expanded, the fluid is cut off at the cylinder and the hose is taken off the shore. The advantage of the speed shore system is that it can be set entirely from above the trench. The disadvantage for rescue use is it does not work well if the walls of the trench aren't vertical or near vertical.

Pneumatic air shores, like those manufactured by Air Shore International and Paratech, come in a wide variety of lengths. Made from lightweight tubular aluminum, the pneumatic shore is quick, strong, and dependable. In general, they are available in lengths from 3 feet to 12 feet and come with a multitude of extensions and attachments, including swivel bases, which allow the shore to be extended at an angle less than horizontal and still be effective.

Paratech Acme Thread Shore

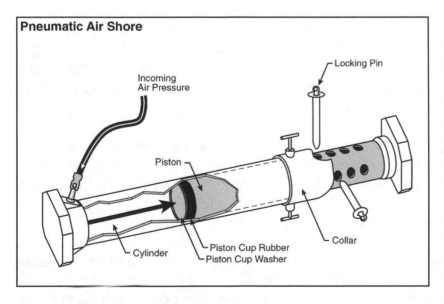
Pneumatic Air Shore
- Locking Pin
- Incoming Air Pressure
- Piston
- Cylinder
- Piston Cup Rubber
- Piston Cup Washer
- Collar

With the exception of the Paratech Acme Thread Strut, which is tightened manually, pneumatic shores all operate under the same principle. The shore is extended by using compressed air at pressures recommended by the manufacturer. After extension, the shore either locks by itself, or is manually locked to prevent a collapse under load. The main disadvantage to pneumatic shores is the number of shores required to maintain an effective cache, and the cost for their acquisition.

TIP: Most shores are only rated for 400 pounds of lateral force when installed. The bottom line is, don't stand on them. They are not steps, so use the ladder.

TOOLS AND APPLIANCES

The variation in tools and appliances required to successfully complete a trench rescue is as encompassing as "give me whatever you've got." All kidding aside, if you would normally find it on a construction site, you can bet it will be needed in a trench rescue.

Ultimately, when we are dealing with the movement of dirt, the **shovel** becomes one of the most important tools we can have during a trench rescue. In the initial stages of a collapse operation, shovels will be needed to move the spoil pile and flatten the area around the trench lip. This will allow the ground pads to lie flat, so as not to create an additional trip hazard on the scene. In addition, you may have a case where a worker is partially buried and the shovel can be given to the worker to begin their own self-rescue efforts. In my experience, most victims that are conscious and trapped in a trench have no problem with self-rescue.

While the shovel might work well at the top of a trench, it has very little value in the trench. The **entrenching tool** is a small collapsible version of the larger shovel, which is designed to be used in situations where room is limited and a shovel is too big. It also gives the rescuer better feel if they are digging in or around a victim.

Remember, digging operations to remove trapped persons should begin as soon as possible after the protective system is in place. That means you will be working around shoring systems and other tight, congested places. Here the entrenching tool earns its money. It will not move a lot of dirt but is great to have when you cannot use a larger shovel and still have to move dirt.

The next most important item to have on the trench collapse site is **the hammer**. You can never have too many hammers. I'm not talking about those little pee wee hammers that you find in discount stores, I'm talking real 20, 22, and 24 oz. framing hammers that will drive a 16-penny duplex nail in three hits. As emphasized earlier in the outline, make sure you not only have the right size hammer, but also give consideration to who is swinging the working end of it. That's when the "right person for the right job" part starts to pay off. Remember, "right person for the right job."

Since trench rescue operations for the most part consist of temporary protective systems, the **duplex nails** we use to connect wood components are designed to be easily removed. The duplex nail can be taken out because it has two shoulders on it, one of which is not supposed to be driven completely flush with the wood. This affords the claw end of the hammer a place to get a bite for removing

Duplex nail being used as toenail for shore

the nail after the operation is over. If you think saving lumber isn't important, it is because you haven't bought any lately. Wood is not cheap and for that reason you should try to reuse all that you can.

The **chain saw** is the saw of choice for rescue operations involving timber shoring. That's right, big saws that will cut your leg off in half a second. Keep in mind that we are not building furniture for sale, we are putting together a protective system that has its emphasis on function and safety, not level and straight. However, a word of caution is necessary at this point, for you must always be alert for the most feared individual on earth…the firefighter with a chain saw. Be careful, you don't get too many second chances with this type of cutting tool.

The ventilation fan can be tied to a piece of lumber to assist in direction of airflow in the trench

Ventilation equipment used in trench rescue is normally the tried and true fire department smoke ejector. Used on the windward side, blowing in the trench, it affords an adequate flow of fresh air in the trench, and when it is hot, provides some relief to the rescuers in the hole. Just keep in mind that ventilation is not always called for unless you have an atmospheric problem. If it is cold outside, all you may be doing is freezing the rescuer and victim. It would be a sad thing for you to spend six hours rescuing a trapped worker only to have him injured from rescuer-induced hypothermia. Use ventilation only when appropriate.

TIP: Tie a few pieces of 12-inch fire line tape to the exhaust side of the fan. The blowing tape will provide a visual indicator that the fan is working. Frequently the noise level on a trench rescue is too loud to notice if the fan isn't operating.

Ladders are another form of equipment that can be used for a multitude of different purposes during a trench collapse. First, you can place a ladder in the trench and ask the victim to crawl by themselves. You would be surprised how fast some people can get out of a trench when you explain to them the complex protective system that has to be set before you can enter. I have seen people with a broken leg crawl up a ladder when they were frightened enough.

Secondary to victim escape is the requirement for ladder egress in trenches over 4-feet deep. As stated earlier in the outline, egress ladders need to be placed so that workers do not have to travel more than 25 feet to get out of the trench regardless of their location. More importantly, you want to provide two points of egress and ingress in the trench for rescuer safety.

Ladders can also be used to span the trench opening and to provide a base for lifting operations over the trench. In addition, ladders can be used as wales when assembled with a 2" x 12" board. While this is not the preferred method to span a trench opening, it will work in a pinch.

TIP: Use only fire-service grade ground ladders for make shift wales. Normal ladders will not withstand the lateral pressures exerted by the shores. Even so, don't exclude the fact that the ladder may be damaged. If damage is a consideration, then get yourself a 6" x 6" or 8" x 8" wood wale.

Dewatering devices are absolutely necessary for the control of water from both ground seepage and the rainwater runoff. Excess water in the trench not only creates an uncomfortable environment in which to work, but it also deteriorates the trench floor and toe if it is allowed to stand.

Mud hog pump for dewatering

Large diaphragm pumps, affectionately called mud hogs, are great low volume dewatering devices that hold up to the rigors of even the worst trench scene. These pumps can be gas, air, or battery operated with each having its own advantages and disadvantages.

TIP: Don't wait for it to rain to look for a dewatering device. In addition, have two of them on site because one of them will inevitably be broken down and not start.

SUGGESTED HAND AND MISCELLANEOUS TOOLS LIST
FOR TRENCH RESCUE

Shovels
Flat
Pointed
"D" Handled
Post Hole
Entrenching

Hammers
Nailing
Large sledge
Small sledge

Saws
Chain
Circular
Hand
Extra blades and chairs
Repair kits
Saw horses

Miscellaneous

Squares	Pickets	**Cribbing**
Pencils	Can of paint	2" x 4"
Paper	Wedges 6" x 6" and 4" x 4"	4" x 4"
Tool ropes (lots)	Ventilation fans	4" x 6"
Tool belts	Fan Duct	6" x 6"
Nail pouches	Power cords	8" x 8"
Nail (duplex)	Generators	
Nail bars	Pike poles	
Tape measures	Air monitor	
Road cones	Buckets	
Barricade tape	Hydraulic rams	

Note: For teaching logistics list, see Appendix Two.

Hand tools for trench rescue

Summary

A trench rescue site is a construction area. Therefore, most types of equipment needed for normal construction are also needed on a trench rescue. Several items that are a necessity are regular shovels and entrenching tools. Hammers for trench rescue should be construction grade, which is a very sturdy and weighted hammer. Because of this, make sure you have the right person swinging the hammer.

Ladders have a multitude of purposes in trench rescue. Whether you use them for wales, lifting platforms, or egress make sure they are of fire service grade. Dewatering equipment will always have to be available on scene. While the centrifugal pump will move more water, the diaphragm pump is usually preferred because of the potential for debris in the water.

Ground pads are used to help distribute the weight of rescuers and other equipment around the trench lip. They can be any size, although typically they are 4' x 8' or 2" x 6", of plywood or lumber, respectively. All ground pads should be installed from a safe area by having the rescuers clear the area in front of the ground pad and then push it forward in front of them.

Sheeting is not generally considered a part of the protective system, and in the case of most trench rescues, it serves to help hold back running debris. Used as an attachment point for the strongback, they can be either purchased or homemade. The strongback is that part of the protective system that transfers energy vertically along the panel and into the trench wall. They are generally made of 2" x 12" pieces of lumber.

Shoring used in trench rescue is that part of the protective system that transfers weight from one side of the trench to the other. Shores can be made of wood or metal and installed manually, hydraulically, or pneumatically.

Questions (Answers and Discussion on page 249)

1. Ropes that are used to set and adjust panels are best attached to the top of the panels.

    a. True
    b. False

2. If 4 x 8 ground pads will not fit on the spoil pile side of the trench you should:

    a. Not worry about it as that side is not generally a problem
    b. Place 2 x 12s on that side
    c. Compact the dirt on that side and use it as a platform
    d. Not work from that side of the trench

3. The use of ground pads is primarily for the distribution of vertical force caused by people standing around the trench lip. The main disadvantage to using ground pads is:

    a. They take time to install
    b. They can cover cracks and separations in the soil
    c. They are expensive and it takes a lot of sheets
    d. They are difficult to install

4. The shores and installed upright in a lumber trench are the only part of the protective system considered by OSHA for compliance.

   a. True
   b. False

5. A duplex nail is designed to be driven entirely into the wood.

   a. True
   b. False

6. When using a ladder as a makeshift wale you should make certain:

   a. It is long enough to span the opening
   b. It is of fire service grade
   c. It is expendable, as damage is possible
   d. All of the above

7. When digging in the trench the preferred tool is the:

   a. Shovel
   b. Entrenching toll
   c. Bucket
   d. Gardening trowel

8. Ladders used in trench rescue can be used as:

   a. Egress devices
   b. Lifting platforms
   c. Angled platforms to effect victim removal
   d. A method to span an slough in the trench wall
   e. All of the above

9. The centrifugal pump can move more water than a diaphragm pump and therefore is the preferred type de-watering device.

   a. True
   b. False

UNIT TWELVE

AIR BAGS FOR TRENCH RESCUE

TERMINAL OBJECTIVE

Understand the proper application of air bags
as used for trench rescue operations.

ENABLING OBJECTIVES

The student will be able to:

- Describe the use and application of high-pressure air bags for trench operations

- Describe the use and application of low-pressure air bags for trench operations

- List the advantages and disadvantages of high-and low-pressure air bags

- Specify a method for determining high-and low-pressure air bag lifting capacity

- Explain the construction features of high-and low-pressure air bags

- Describe the proper procedure for using cribbing to provide stabilization during lifting operations

The air bag, as used for lifting objects, was introduced in the seventies, and is one of the very best tools developed and adapted for the rescue business. Using air supplied to a balloon-like vessel, the pneumatic air bag can perform many functions at a trench rescue. Some of these uses involve lifting heavy objects like pipes, excavators, steel panels, and concrete distribution boxes—objects that workers seem to sometimes find themselves under, instead of beside. In addition, air bags can be used to fill voids created by sloughs and other types of trench collapses.

Keep in mind that this text is not a comprehensive guide to pneumatic air bags. Our task will be to explain their construction, operating principles, and specific applications for trench rescue activities. This being said, it is a plus that no matter what the specific application of air bags, the principles of operation always remain the same.

AIR BAGS IN GENERAL

Air bags can be purchased as low, medium, or high-pressure systems. The amount of air pressure supplied to make the lift determines the classification of the bag. Low-pressure bags use 7 psi, medium pressure 22 psi, and high-pressure 80 to 120 psi. For the most part, you will not see too many sets of medium-pressure air bags, and for that reason, we will concentrate our efforts on learning the low and high-pressure systems.

Generally speaking, the low-pressure air bag will only lift limited capacities, as compared to high-pressure bags, but will lift the objects higher. High-pressure air bags will lift a greater amount than low-pressure bags, but will not lift the object nearly as high. Using the proper tool for the job is the key to successful air bag operations at a trench rescue.

HOW AIR BAGS WORK

All air bags are supplied a volume of air under pressure from a remote air source. This source is usually a self-contained breathing apparatus bottle, but could be an air compressor, or even a hand pump. In most applications, the air

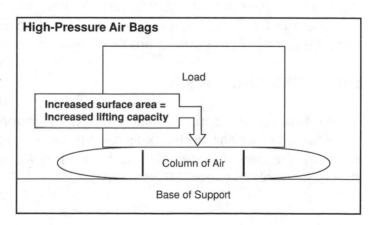

High-Pressure Air Bags

Load

Increased surface area = Increased lifting capacity

Column of Air

Base of Support

source will supply a regulator that reduces the pressure and supplies a controller. The controller is the distribution point for multiple airlines, and dictates the rate of flow to the individual bags.

All air bags use the same principle to effect the lift of an object. The effectiveness of air bags is tied to the compressibility of air, as divided over the inside surface of the air bag. Basically, the lifting capacity of any air bag is limited to the amount of bag surface area that can contact the object, multiplied by the operating pressure supplied to the bag.

Air bags are rated at a maximum and minimum lift. This takes into consideration the surface area of the bag that is in contact with the object. As the bag inflates, it becomes football shaped and loses surface contact with the object. As this occurs, the lifting capacity of the bag diminishes proportionally.

To best illustrate this concept, pull out a set of high-pressure air bags and measure the length and width of one of the bags. Take these measurements and multiply them together, then multiply the result by the operating pressure as recommended by the manufacturer. Now compare your figure with the maximum lifting capacity as indicated on the bag. Your figure and the manufacturer's should be about the same. If not, maybe math is not your best subject.

Now that you have a handle on this concept, let's apply it to a rescue-lifting scenario. The object you want to lift has a surface area of 6 inches x 6 inches, and weighs 6,000 lbs. You have selected a 40-ton air bag to manage the lift. The 6" x 6" surface area of the object to be lifted equals 36 square inches and the operating pressure of your air bag is 120 psi. By multiplying the 36 square inches by the 120 psi operating pressure, you obtain a lifting capacity of 4,320 lbs. Since 4,320 does not equal 6,000, you have a problem. At this point, it does not matter if your bag can lift a trillion pounds; you do not have the surface area to effect the lift. The moral to this story is to find yourself a lifting point on the object that has more surface area.

HIGH-PRESSURE BAGS

High-pressure bags consist of rubber or neoprene material reinforced with steel bands or kevlar, and usually have a coarse surface to increase the friction between the bag and the lifting surface. They are operated with an air system that can supply between 80 and 120 psi.

Stacked high-pressure air bags used on platform over trench

As stated earlier, the drawback to high-pressure bags is that they do not lift objects very high. To offset this limitation you can stack them. The rule of thumb for stacking high-pressure air bags is to stack only two, and to always put the largest capacity bag on the bottom. Just keep in mind, the total capacity of your lift is limited to the lowest capacity of the stacked bags.

High-pressure bags, while very durable, are not field repairable should they develop a hole. Remember this when you are providing the bag protection from sharp objects. If the air bag blows under pressure, you will run a mile before you turn around to see what created the noise. Trust me...I've been there.

TIP: While stacking bags will increase the height that an object can be lifted, it will not increase the capacity of the lift. To achieve a higher capacity lift, put the air bags beside each other (i.e. utilize more surface area of the object to be lifted) and inflate them simultaneously.

LOW-PRESSURE BAGS

Low-pressure air bags are flexible rubber bags, used primarily in trench rescue to fill voids in trench walls, although they can be used to lift some objects. Operated at 7 to 12 psi, low-pressure bags will lift objects higher than high-pressure air bags, although they will not lift nearly the amount of weight. The nice thing is that if low-pressure bags get a hole in them, you can just plug it with your finger. For the most part, this makes them field repairable. Additionally, consider

Low-pressure air bag for lifting

that the average 16-ton air bag may require as much as 250 cubic feet of air to effect a lift.

Low-pressure bags can also be used to lift objects by placing them on a platform outside the trench, and above the object that needs to be lifted. By taking a rigging strap and wrapping it around the bag, then around the object to be lifted, the lift is effected as the bag is inflated. In this scenario, make sure your lifting platform is substantial enough to hold the weight of the load until it can be cribbed.

In a different type of application the low-pressure bag can be used to fill a void. For instance, with a rotational failure there will be a section of the wall that has fallen in the trench. To transfer the forces from one side of the trench to the other, this void must be filled. A large low-pressure air bag can be inflated between the wale that spans the opening and the existing wall for this purpose.

Low-pressure air bag filling void

TIP: Keep in mind that the higher any bag lifts, the more unstable the lifted object may become. There must be a constant evaluation of the lifted object's center of gravity compared to its stability. Always remember to: ***Lift an inch —crib an inch!***

CRIBBING

No section on the lifting of objects would be complete without a description of the proper use of cribbing. This is because during each lift, the object being lifted should never be more than an inch from a substantial cribbing system. The object should continue to be supported by the bag after the lift is accomplished.

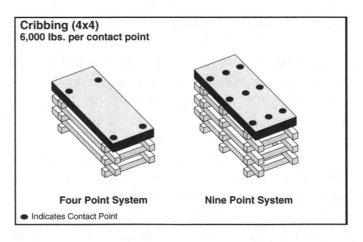

Cribbing (4x4)
6,000 lbs. per contact point

Four Point System   **Nine Point System**
● Indicates Contact Point

Cribbing material for trench rescue activities is usually cut out of 2" x 4" and 4" x 4" pieces of lumber. The latter is what the boxes are usually constructed from, with the former used to take up odd spaces. Generally, cribbing is cut to standard 12, 18, or 24-inch lengths. Don't go all out buying wood for cribbing. What I mean is that this is not the place to invest your money, since just about any construction grade lumber will work. Keep in mind that over time the wood will lose its natural moisture content and can become brittle or split under a load.

THE BOX CRIB SYSTEM

There are three variations of the box crib system—the 4 point, 9 point, and the full box crib. Each system is so named by the number of transfer points in the cribbing system. Thus, a 4-point box crib system has two timbers for each layer, but a total of four places that each piece of lumber crosses. A 9-point box crib has nine transfer points and three pieces of cribbing on each layer of the system. A full box crib system would be solid on all layers—a very strong system, but certainly lumber intensive, and not usually called for in trench operations. When using the box crib as a lifting platform, the top layer should always be solid.

The general rule of thumb for box cribbing is that each point of contact will support a standard amount of weight, depending on the size of the lumber. A 4" x 4" will support about 6,000 pounds per point, and the 6" x 6" crib about 15,000 pounds per point. By adding the number of points together, you can determine the total capacity of the

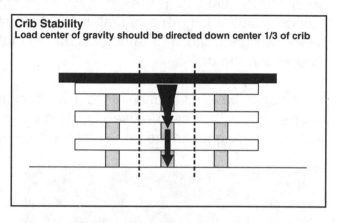

Crib Stability
Load center of gravity should be directed down center 1/3 of crib

box crib. Just remember that since the weight supported may be more concentrated when on the box crib (less surface area), consider whether the ground upon which the box crib rests is substantial enough to hold the load. This is especially true in the bottom of a wet trench.

Box crib for lifting support

All box crib systems are made by stacking the timbers in alternating rows, each row at a right angle to the previous row. Each row should be stacked and placed at a distance somewhat less than the total length of the piece of timber, thus, not creating too much overhang, but always situated past the outside edge of the piece directly under it. The height to which the box crib should be built is no more than three times the diameter of the base. Therefore, a 2' x 2' box crib should be no higher than six feet tall; however, it may be less depending on the lifted object's center of gravity and the stability of the ground supporting the box crib.

WEDGES

Wedges are cut pieces of lumber that form an inclined plane. The plane of the wedge makes it adjustable depending on the space it needs to occupy. Used in this fashion the wedge takes up the space between the object being lifted and the

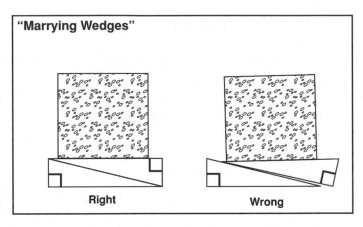

box crib, until a full piece of timber will fit under the object. Thus, if something happens during the lift, the object can only fall the limited space between the wedge and the object.

Wedges have a variety of uses at a trench rescue. As indicated, by being an inclined plane they can be used to tighten objects and take up spaces between waler and uprights. In the timber trench they are the primary method to tighten the shores to the uprights. The bottom line is that you can never have too many wedges.

Wedges used to create contact points with trench panel

Wedges used to fill the void between the wale and strongback

Summary

Air bags for trench rescue can be used for many types of applications. Most of the time air bags are used for lifting objects and filling a void. High-pressure air bags are used to lift heavy objects but can only lift them a limited height. Low-pressure air bags can only lift a limited capacity but can lift that capacity much higher. The lifting capacity of all air bags is limited to the pressure supplied to the bag and the lifting area of the bag with regard to the surface area of the object being lifted.

Box cribbing is used to support the object being lifted both during and after the lift is completed. Box cribs can be made as four points, nine points, or full box. The space available and the weight of the lifted object determine the type of box crib necessary.

Wedges are used during the lift to take up the space between the box crib and the object until a full piece of lumber can be used. The rule of thumb for cribbing is "lift and inch—crib and inch."

Questions (Answers and Discussion on page 251)

1. When lifting in the trench environment you must consider which of the following:

   1. Weight of the object
   2. Height of the object
   3. Type of object you are lifting
   4. Water conditions in the trench
   5. Type of lifting device available

   a. 1,2,3
   b. 2,4,5
   c. 1,2,3,5
   d. 1,2,3,4,5

2. High-pressure air bags offer the following advantages:

   a. Higher weight lifting capacities
   b. Will lift the object higher than low-pressure bags
   c. Offers puncture resistant construction
   d. None of the above
   e. Both a and c are correct

3. Low-pressure air bags are usually field repairable.

   a. True
   b. False

4. The ability of an air bag to effect a lift is a function of:

    a. The surface area of the object being lifted
    b. The pressure supplied to the bag
    c. The surface area of the bag
    d. All of the above

5. A four point (4" x 4") box crib will theoretically support

    a. 15,000 pounds
    b. 30,000 pounds
    c. 23,000 pounds
    d. 24,000 pounds

UNIT THIRTEEN

TRENCH RESCUE ASSESSMENT

TERMINAL OBJECTIVE

Be able to formulate a strategic plan based on the determination of facts surrounding the trench emergency.

ENABLING OBJECTIVES

The student will be able to:

- Identify those factors that would be pertinent in formulating a trench emergency plan before arriving at the scene

- Describe the appropriate questions to ask about the event after arrival at the scene

- Explain factors to be considered during the incident

- Summarize the steps to consider when looking for buried victims

After the preparation phase of trench rescue, probably none of your efforts will be as important as assessment. Assessment is the foundation on which you build your decision-making platform. When completed, it helps you determine a set of guidelines for action. Basically, the assessment is nothing more than a tried and true situational size-up. Size-up is taking the time to figure out what has happened, analyzing the information, and then providing a coordinated plan to address the situation.

Assessment can be broken down into three time periods: from the time of alarm until you arrive on scene, upon your arrival on scene, and continuously during the operation. If you are an emergency service provider, most of these considerations come as a natural part of the rescue response. However, here are a few considerations that are specific to trench rescue operations.

TIME OF ALARM

At the time of the alarm you will begin the process of gathering information. This information will come from alarm data that is supplied to the dispatcher at the time the call is received. The typical call will reference a collapse of some sort and information will be very sketchy. Usually, you will get very little detailed information. The reason is that trench rescue is a specialty call that doesn't happen very often, and the initial questioning of the caller by the dispatcher will more than likely be less than adequate. It will then be up to you to question and prompt the dispatcher for additional information while you are en route. Some of the questions that may be appropriate are:

What has happened?

Have the dispatcher call back and keep the complainant on the line. This will be critical if you are having trouble finding the accident scene, but at any rate, someone will be available if you have additional questions.

Why was the excavation work being done?

If the hole is for a small utility line it is one thing. If it is for a large storm drain pipe it is quite another. This type of information will give you an

What type of work is being done?
Big pipes mean deep trenches

indication as to how large the excavation was before it collapsed. Consideration may be given to acquiring additional resources at this point.

Is the victim(s) completely buried?

This is the step in the information process that allows you to begin your risk assessment. It also aids in determining the amount of resources that will be required to get the job done.

Is the situation a trench collapse or some other form of injury in the trench?

As we mentioned earlier, most trench emergencies will be for something other than a collapse. If the victim is injured, but not buried, you are dealing with a very different type of call.

Will I have access problems for the equipment and rescue personnel?

Generally, new construction areas don't have an established road network. This could mean a delay in getting resources to the scene. All of the equipment needed may have to be carried to the site if you can't get your apparatus close to the collapse.

Assessment factors should include weather conditions, traffic, time of day, and how close traffic is to the trench collapse

How is the weather, and can I expect it to change if this becomes a long-term operation?

The collapse itself may have been weather induced, but in any case, consideration may have to be given to changing weather conditions. For instance, you may need de-watering equipment if it is raining or if a storm is expected. If it is going to get hot during the rescue you may need more personnel than would normally be necessary. Another area that is closely associated with the weather is the need for lighting during overcast conditions or should the situation extend into the night.

TIP: Have all resources that are responding to the scene stage at a remote location until your assessment is complete. A good assessment is hard to do with a bunch of action oriented folks on the scene. Trust me, rescuers are not a patient bunch!

ARRIVAL AT THE SCENE

Upon arrival at the scene you will be expected to finish gathering your information and develop a plan for mitigating the incident. This means taking the basic information you have been able to gather, adding to it what you can subsequently glean from witnesses and assimilating all of it with what you actually can see with your own eyes. Below are some additional on-scene considerations.

Who is in charge and what has happened?

Seek out and question the competent person on site. This person will give you information as to the original depth and width of the trench and what type of protective system was, or is, in place (if any). You will also want to find out what the victim was doing at the time of collapse, and where he was last seen.

Is there a language barrier?

Nothing can be as frustrating as trying to handle an emergency and not being able to communicate with the person who can give you the necessary information to be successful. If there is a language barrier, find someone to interpret.

Based on equipment limitations, is the collapse within your scope of operations?

Remember that your equipment has limitations. Generally, if the trench is deeper than 15 feet, or if there has been a massive cave-in, commercial techniques for stabilization will be necessary.

What are the injury problems?

How serious are the victim(s) injuries and do I have discretionary time in providing treatment. For instance, will the extrication process take so long that the victim may die, or do I just hook him with a pike pole and pull him out? While we want to always maintain good C spine immobilization for our victim, whether or not he ends up paralyzed may not be a consideration if he will die because it took us too long to get him out.

What is victim survivability profile?

This one is another no brainer. If your victim is buried, and there is no chance they jumped in the end of a pipe for protection, you're most likely dealing with a recovery. Just remember a recovery is no longer an emergency. Don't get your people hurt!

What type of protective system is/was in place?

This will be important, because if the original protective system failed, you have a big problem. It may well be that you have to remove what was in place and then start over, or try to figure out a method to stabilize the existing system. At this point, if you are dealing with a fatality, call OSHA or other professional engineers for help.

DURING THE EMERGENCY

Keep in mind that everything done on a trench collapse scene makes it a new scene. These situations are so dynamic that constant evaluation for changing conditions is of paramount importance. Constant evaluation will help you anticipate potential problems in order to remain proactive and not become reactive.

All of these factors, and many others, will need to be addressed if you are going to be successful at a trench collapse. If there is any part of your operation that you must take your time completing, then this is the one. Good information will lead to a good plan of attack. Don't make a move until you have fully assessed the situation, or you may well find yourself looking at the FAILURE acronym to see where you went wrong.

FACTORS TO CONSIDER WHEN LOOKING FOR BURIED VICTIMS

For obvious reasons, the first place you would want to look for, or would expect to find a victim of a trench collapse, is at the end of the pipe string. This area represents the last location where work was taking place. Other information such as the depth of the trench may be determined by looking at the engineer's flagstick. Normally, the flagstick will be marked with the trench depth and grade information. You may also be able to look at the orientation of laser targets if they have not been moved. However, more than likely when you get the call for a collapse it will be noticeable, and the competent person will have a good idea of the general area in which the victim was last seen.

If the collapse has occurred at the end of a pipe string, you should definitely listen for sound in the pipes. It may well be that the victim was able to get their head and chest in the exposed pipe before the collapse. Go back to the point of origination, or nearest entry point for the pipe, and listen for sounds from the victim. Keep in mind that if you enter the pipe for rescue purposes, it is a confined space and additional considerations for this type of rescue are necessary.

More difficult is actually finding a buried victim when his location is in question. One of the things you can do is use some of the information you gathered during the assessment. Consider what the victim was doing at the time of the collapse, and then look for indicators of this type of rescue. For instance, if they were sighting the hole for depth, you may well find the victim at the bottom of a grade pole. In addition, paint and grease buckets may have been located on top of the trench, well within reach of the worker. The same can be said for tools and other equipment.

Exposed limbs are, for obvious reasons, a good indication of victim location. It is only mentioned here because caution has to be taken when assuming victim location based on exposed body parts. The exposed limb may well not be in normal orientation for the human body. Be careful when digging around victims until you are sure of the head and chest location.

TIP: By far, the best indicators of victim location are nearby drink containers. Refreshment items are most always well within reach of workers.

At the conclusion of your assessment, you will be ready to develop a rescue plan for the incident. This marks the official start of the extrication effort. Prior to any activity taking place on the scene, you must make sure all rescue team members are briefed concerning your tactics and strategies. This meeting is generally referred to as a pre-entry briefing and should detail the identification of all known hazards, emergency contact signal, the command structure, radio frequencies, and tactical objectives.

Summary

As with any emergency scene, there are many factors that need to be considered when doing a trench rescue assessment. These factors can be broken down as things to consider prior to arrival, assessment factors after arrival, and the ongoing assessment considerations.

Assessment information prior to arrival will help you determine the equipment necessary to mitigate the incident, while others will help you do your risk benefit analysis. For instance, the location of the trench collapse, the type of work being done, access problems, and the current weather are very important considerations prior to arrival. The condition of the victim, type of collapse, and future weather are all considerations after arrival. On-going assessment considerations are the integrity of the stabilization system and other on-scene activities.

Factors to consider when looking for a buried victim include what the victim was doing at the time of the collapse and any information from persons who may have witnessed the event. Other high probability areas include the pipe string or near an engineering flagstick. Other than exposed limbs, the location of personal items like sweat towels and drink containers are good indicators of where victims might be buried.

Questions (Answers and Discussion on page 252)

1. Clues as to the location of a buried victim include:

   a. Grade poles
   b. Tools
   c. Drink cups and other food containers
   d. Shovels or other digging equipment
   e. All of the above

2. The determination of facts and conditions that led to the collapse of a trench is called a:

   a. Size-up
   b. Discovery of facts
   c. Determination of events
   d. Pre-entry decision making matrix

3. Which of the following are important considerations in the size-up of a trench rescue:

   a. What has happened?
   b. What was the work being done at the time of collapse?
   c. Is the victim completely buried?
   d. Are there access problems for equipment and manpower?
   e. All of the above are correct

4. Which of the following are important considerations after arrival on the scene of a trench rescue:

   a. What has happened?
   b. What was the work being done at the time of collapse?
   c. Is the victim completely buried?
   d. Are there access problems for equipment and manpower?
   e. All of the above are correct

UNIT FOURTEEN

HAZARD CONTROL

TERMINAL OBJECTIVE

Understand the various hazard types, categories, and phases of control that will be found at the trench rescue scene.

ENABLING OBJECTIVES

The student will be able to:

- Describe the various types of hazards that can be found at a trench rescue

- Identify the five hazard control categories

- Explain the phases of hazard control at a trench emergency

Your plan to handle the emergency is now at the point where real and potential hazards will need to be addressed. Hazard control is really the last phase of assessment, although it is so important that it warrants its own phase.

Of the many different hazards that can affect your operation, all of them can be categorized into one of two types: The hazards you are able to control and the hazards you should leave alone.

Hazards that can easily be controlled, and that are within the expertise of the rescuer, need attention before any deployment of your personnel. These types of hazards could be the location of vehicles, trip hazards, spoil pile movement, and supporting existing utilities, etc. Examples of those hazards that should be addressed by someone else are electricity and gas. The rule of thumb is that if a professional discipline has been established for the hazard, it is a good indication you are not qualified to handle it. Call a professional if there is any question or if the situation is beyond your training and abilities.

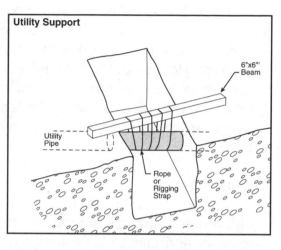

Unsupported water line

TIP: Support all known utilities that cross the excavation before or during the process of building your protective system. Having an unsupported water line break in the middle of your operation could get really ugly.

HAZARD CONTROL CATEGORIES

Hazards can be broken down into five categories: mechanical, chemical, man-made, electrical, and water.

Mechanical

Machines and other entrapping mechanisms could be a danger to rescuers. Just as in Confined Space "lock out tag out" procedures, make sure you bring everything to a zero mechanical state. Bringing something to zero mechanical state means eliminating all possibility that any activation could take place. Examples include, taking the keys out of machinery, locking out electrical devices, and removing machines that are in the collapse zone.

Chemical

As was discussed earlier, you should always assume that something could have been unearthed during digging operations. In addition, a worker may have carried a chemical in the trench for use during intended work activities. Examples include gasoline for saws, solvents for cleaning, and glue for making pipe connections. The lesson here is to never assume anything is safe, and always monitor the atmosphere.

Gas line in trench

Man-made

Man-made hazards are usually the reason that you are there to begin with. These are normally all of the things we do as a part of our normal work that can be hazardous at some point in the operation. Examples of man-made hazards are spoil pile location, equipment location at the scene, and tripping.

> **TIP:** When considering shutting down heavy equipment that is running, consider if it is attached to anything. If you elect not to move the equipment, keep the operator on scene, but in all cases take the key.

Electrical

As a firefighter it has always been a philosophy of mine that I won't mess with electricity, and conversely, I won't expect an electrician to fight fires. I'm not referring to shutting off breakers here, but more along the line of poles, cables, and transmission box hazards. Control of electricity is best left to the people that do it for a living, not us part time hazard control experts. Also watch out for telephone lines. The voltage that is carried in telephone lines will knock the fillings right out of your teeth. As a practice, determine all util-

Mark all utilities and make personnel aware of their location prior to rescue work

ity locations before digging in a collapsed area. This can be as simple as calling the local utility marking service company located in most areas.

Water

Water can be a hazard on the scene of your collapse by already being in the ground or as the result of rain. If rain is imminent, start thinking about building a cover for the trench and establishing a method to divert incoming rain. The bottom line is to get existing water out of the trench and do your best to see that no additional water gets in it. Have dewatering equipment on site.

TIP: Have a Hazardous Materials Team as part of your initial response matrix for atmospheric monitoring purposes. In addition, have personal contact numbers for each professional discipline that could represent a collapse scene hazard.

HAZARD CONTROL PHASES

After the identification of all hazards on the scene of the collapse, you will want to turn your attention to addressing them in some sort of logical order. For obvious reasons, the most hazardous might warrant first consideration. However, don't blindly address what appears to be the most hazardous element without consideration for those hazards which may not be as readily apparent.

Just as in other rescue training in which you may have participated, there is standard nomenclature used to describe the various areas of the emergency scene. In Haz-Mat you remember it as "hot," "warm," and "cold" zones. In vehicle rescue you may have been taught about the "inner" and "outer" circles. Trench rescue incident scene management could use these terms or the "general area" and "rescue area" to define and establish boundaries of operation.

The **general area** would be the area that is not in the immediate vicinity of the extrication effort. Hazard control in this area would entail a large overview of the scene and, under normal circumstances, should begin first. The activities that would normally take place in this area are: staging, cutting, logistical support, rehab, and vehicle parking.

TIP: In general, the minimum safe area to establish around a trench incident is 300 feet.

The **rescue area** would be located immediately surrounding the rescue site. This is a very small area encompassing the rescue effort, and hazard control would only be considered after the general area was completed. Extrication efforts/activities such as air supply, panel team, shoring personnel, and the safety officer will be established within this area.

TIP: Your personnel can become a hazard if too many of them end up in the rescue area. Keep all personnel not actively involved in the rescue effort in the staging area.

NOTE: Ventilation as a hazard control technique will be covered in Unit 15— Atmospheric Monitoring.

See Unit 19, page 185, for an illustration of trench scene management using the Hot, Warm, and Cold Zone terminology.

Summary

There are many types of hazards found on the scene of a trench collapse. These hazards can be assessed as those in which you have the expertise to control and those best left to a professional. In general, if there is a professional occupation established for a type of utility, then you should probably use them to help you gain control of it.

There are five hazard control categories: mechanical, chemical, man-made, electrical, and water. All of these hazards should be eliminated before beginning any rescue effort. Your local utility marking service can be a big help in this area.

The trench scene can be divided into hazard control zones like hot, warm, and cold. Still another terminology is general area and rescue area. Whichever terminology you choose, make sure all personnel are aware of what takes place in these areas and the hazards located in them.

Questions (Answers and Discussion on page 253)

1. Which of the following is not one of the hazard control categories:

    a. Mechanical
    b. Hydraulics
    c. Electrical
    d. Water
    e. Chemical

2. The control of scene hazards takes place in two phases, the general area and the logistics area.

    a. True
    b. False

3. When considering shutting down heavy equipment at the scene you should consider:

    a. If the equipment is attached to a structure in the trench
    b. Taking the keys out and putting them in your pocket
    c. Keeping the equipment operator at the staging area in case he is needed
    d. If it is safer to move the equipment before shutting it down
    e. All of the above are correct

4. The minimum safe area that should be established around a trench incident is:

    a. 150 feet
    b. 200 feet
    c. 100 feet
    d. 300 feet

5. Knowing the location of all underground utilities is important before beginning the rescue effort. For this reason it is a good idea to:

    a. Include this step as a part of hazard control
    b. Call your local utility marking service
    c. Mark the utilities and make all personnel aware of the locations
    d. All of the above

UNIT FIFTEEN

ATMOSPHERIC MONITORING FOR TRENCH RESCUE

TERMINAL OBJECTIVE

To apply the concepts of proper atmospheric monitoring to trench rescue emergencies as a method of determining existing and potential hazards.

ENABLING OBJECTIVES

The student will be able to:

- Describe various scenarios in which a rescuer could be confronted with an atmospheric problem at a trench rescue

- Recite the definition of a permit required confined space

- Understand the definitions that apply to atmospheric monitoring

- Specify the various considerations for monitoring the atmosphere in and around a trench

- Describe the various action guidelines as they apply to oxygen, flammability, and toxicity

- Summarize the nine rules of atmospheric monitoring at a trench rescue

- Describe the use of ventilation as a hazard control option

At any given time it would be appropriate to ask yourself, why do I need to know about atmospheric monitoring to handle a trench rescue? I would have probably asked this same question some years ago. Unfortunately, in this day and time, our thought process from a hazard recognition perspective needs to be much broader.

A note of caution at this point: this is not a manual on Confined Space Entry and Rescue. For that reason, the amount of atmospheric monitoring material we cover will not be as comprehensive as it could be. Remember, this is a trench book, so don't rely on it as an all-inclusive guide to atmospheric monitoring or hazards associated with atmospheres. This information should be used to make you aware of potential problems and what actions you should take with regard to hazardous atmospheres.

TIP: Check out CONFINED SPACE RESCUE written by Chase N. Sargent for a comprehensive guide to this rescue discipline.

The advent of the Haz Whooper Standard has brought with it a liability for anyone who has, or is trying to, dispose of a hazardous substance. This has generated a whole new set of problems for rescuers who unknowingly get involved with the improper disposal of materials. Keep in mind that it normally costs a lot less to bury something than to properly dispose of it.

How does this affect the procedures used in a trench rescue? you ask. Let's look at three very real scenarios to form the basis of our discussion on atmospheric monitoring.

Scenario one

A worker is operating a chain saw for cutting concrete forms in the bottom of a trench. The five-gallon gas can he is using to refill the saw is also located in the trench. A subsequent collapse covers the worker and crushes the can, allowing the contents to escape.

Scenario two

A worker is using a gas-powered generator to operate lights for a nighttime digging operation. You get a call regarding an unconscious worker in a trench with no collapse. Carbon Monoxide (CO) from the generator is also permeating the trench.

Scenario three

A worker is standing next to an excavation, directing a digging operation. The backhoe operator strikes a hard object in the ground, which happens to be a 55-gallon drum of toxic waste that has been previously buried. When you arrive, both workers are unconscious.

These are three very real and potentially deadly scenarios that could have a very profound impact on the outcome of your rescue. Having a basic under-standing of atmospheric monitoring and the effects of hazardous atmospheres is imperative to successful trench rescue operations.

Confined Space or Trench

OSHA believes there are confined spaces in about 238,853 workplaces, and of the 12.2 million workers employed at these establishments, approximately 1.6 million of them enter 4.8 million <u>permit required</u> confined spaces each year. While a trench is not exactly a confined space by definition, let's examine the definition of a confined space to see if any similarities exist.

Confined Space:

- Large enough and configured so that an employee can bodily enter the space
- Has limited mean of egress for entry or exit
- Is not designed for continual employee occupancy
- Has an actual or potential hazardous atmosphere
- May also have any of the following:

    1. Material with the potential of engulfing the entrant
    2. An internal configuration that could trap or asphyxiate an entrant due to converging walls or sloping and tapered floors
    3. Any other recognized serious safety hazard

Now I'm not the smartest person in the world, but it seems to me there are a lot of similarities between a trench and a confined space. Since atmospheric problems are responsible for the overwhelming majority of deaths in confined spaces, it is probably a good idea for us to take a hard look at the subject as it applies to trench rescue.

Atmospheric monitoring of a
trench prior to operations

Atmospheric Monitoring

The use of air/monitoring and sampling equipment is one of the most important aspects of a confined space or trench rescue operation. During your rescue effort someone on your team, (usually Haz-Mat or support function personnel), should be providing periodic monitoring in and around the trench. Monitoring is used not only to detect the presence of Immediately Dangerous to Life and Health (IDLH) atmospheres, but it can also be utilized as a tactical guide to ventilation of the trench.

Before we go any further, we need to have an understanding of some basic terms that apply to monitoring and sampling. This will enable us to better understand the technical operation of monitoring and the response data we get from a given sample.

Definitions

Alarm settings: An alarm setting is the preset level within a monitor at which the monitor will display a visual alert and sound an audible alarm. Alarm settings are established by the manufacturer and are based on OSHA (Occupational Safety and Health Administration) and NIOSH (National Institute of Occupational Safety and Health) standards for a given product. For confined space we refer to these as "Action Guidelines."

Detection: The act of discovering the presence of a contaminant in a given atmosphere.

Detection Range: A term expressing the unit of measure a monitor uses to detect the vapor for which it was programmed. Combustible Gas Indicators (CGI) usually have a display showing percentage (%) of the Lower Explosive Limit (LEL). Toxic sensors such as Carbon Monoxide or Hydrogen Sulfide units display in parts per million (PPM).

Dusts: Byproducts of solid materials that have been sanded, ground, or crushed.

Explosive limits: A reading or display on a monitor indicating a percentage of gas in air mixture. These readings are known as Upper Explosive Limit (UEL) or Lower Explosive Limit (LEL)

Flammable Range: The percentage of vapor in the air that must be present to sustain combustion should an ignition source be present.

Flash Point: The minimum temperature at which a combustible substance generates enough vapor to form an ignitable mixture with air in the vapor space above itself.

Ignition Temperature: The minimum temperature to which a liquid must be raised in order for combustion to be initiated and sustained.

Immediately Dangerous to Life and Health: (IDLH) Maximum concentration from which a person could escape (in the event of respirator failure) without permanent or escape-impairing effects within 30 minutes.

Lower explosive limit (LEL): The minimum concentration of vapor in the air at which propagation of flame occurs upon contact with a source of ignition, usually expressed as a percentage of gas vapor in the air.

Oxygen Sensor: An electrochemical sealed unit that measures the percentage of oxygen in the air. The sensor has two electrodes, an electrolyte solution and a membrane, which separates the two. As oxygen passes through the membrane, a reaction with the solution and the electrodes produces an electrical current, which causes the sensor to display the percent of oxygen (O_2) found.

Permissible Exposure Limit: (PEL) Average concentration that must not be exceeded during an 8-hour work shift or a 40-hour workweek.

Upper Explosive Limit (UEL): The maximum concentration of vapor in the air at which propagation of flame occurs upon contact with a source of ignition, usually expressed as a percentage of gas vapor in the air.

Monitoring Considerations

Before you can actually monitor a space, there are several things you should consider regarding the atmosphere or potential hazards that exist at a trench rescue operation.

1. **What is the nature of the hazard I am monitoring?** You should know something about the product you are testing for from your previous evaluation, Material Safety Data Sheet (MSDS), or intelligence you gather at the scene. Do you know the upper and lower explosive limits for the particular

product? Is the atmospheric oxygen deficient, which might create interference with instrument response? What's the vapor pressure of the product and what's the outside temperature? Is it likely that this combination will create enough vapors to support ignition? Remember, if a product is producing vapors then the product is "coming after you." You don't have to go after it to be intimate with it and this is a very dangerous situation if not controlled. Finally, is the vapor a health hazard, and is the material lighter or heavier than air? This knowledge provides a clue regarding where the product might lay and how it might move.

2. **Are there sources of electrical interference around?** Electromagnetic fields, high voltage, static electricity, portable radios and things like cellular phones can interfere with your meter readings. Many current instruments offer RF shielding as a feature. Consider this when you operate.

3. **What are the environmental site conditions you are operating in?** Things like temperature, humidity, barometric pressure, elevation, sunlight, particulates, and oxygen concentration have to be considered.

4. **Are there gases and vapors that are interfering with your monitor?** An example of this is the lead in leaded gasoline. This permanently desensitizes the filament of the combustible gas indicator so that it is unable to detect anything! Certain acids and corrosives will eat your monitor and the sensors, rendering the meter useless. Manufacturers supply information about their particular meters, and you should be familiar with yours.

Action Guidelines

In order to tactically use monitored information, you must have action guidelines established. In most instances these guidelines are outlined in OSHA 1910.146 (the confined space standard) and should be incorporated into your SOPs.

Action guidelines indicate that you should take some specific action(s) when monitor readings reach certain levels. As you begin to look at action or alarm guidelines established by different agencies such as EPA (Environmental Protection Agency), OSHA, and others, you will find that these guidelines differ. Some are specific for hazardous materials events, some are for hazardous waste sites, and some relate to confined spaces. The action guidelines presented here are for confined space and trench rescue operations and should not be confused with other operations. Action guidelines are exactly that—guidelines.

There are usually preset alarms on your monitors that will alert you when a certain level, parameter, or product is detected.

Atmosphere	Level	Action	Monitor
Combustible/ flammable gas	10% of the LEL	If outside the space, correct atmosphere. If inside the space, begin exit.	Alarms both visually and audibly
Oxygen	Less than 19.5% or greater than 23.5%	If outside space, determine problem and correct. If inside space, begin to exit.	Alarms both visually and audibly
Toxicity	CO - 35 ppm H2S - 10 ppm	If outside the space, determine cause of problem and correct. If inside the space, begin to exit.	Alarms both visually and audibly

General Monitoring Guidelines

The following are my friend Chase Sargent's Nine Rules for Atmospheric Monitoring.

Rule one: Monitor in order. There is an order or hierarchy utilized in monitoring and detection that we will use for rescue operations. This differs from hazardous materials in some ways, since we do not monitor for radiological hazards unless there is a high suspicion of the presence of such materials. Generally, however, the order of monitoring will be as follows:

1. Oxygen
2. Flammability/Combustibility
3. Toxicity

(Note: For all of you Haz-Mat gurus that are saying PH should have been listed first, refer to the opening part of this chapter that deals with what type of manual it is.)

As we further discuss specific monitoring considerations, you will understand why this order has been established. This process should be done each time you monitor.

Rule two: Always monitor at multiple levels in the trench: Any flammable vapor or gas, even below the LEL, can be hazardous to you. Mixtures of gases can accumulate at different locations in the trench depending on individual gas vapor densities. Vapor density is nothing more than the nature of a gas to rise or fall! Air has a vapor density of 1.0; gases and vapors with a vapor density of less than 1.0 will rise, while those with a vapor density greater than 1.0 will sink.

Always monitor at three levels within any space. Different products have different vapor densities, which means that they will locate in a trench at different levels. If for some reason you only measure the bottom one third of the trench, you may miss at least two or three different products that might be present.

As an example, Methane is typically lighter than air and will leave the trench. Carbon Monoxide (CO) is about the same density as air and it will tend to locate mid-trench or diversify with the air. Hydrogen Sulfide (H2S) is heavier than air and will tend to find its way to the bottom of the trench. Sampling technique is important.

Rule three: Know your monitor's limitations: Monitors have limitations. Not just technological limitations, but limitations on how accurate they are when detecting certain substances. You must know what limitations there are to your monitoring capabilities, and how certain readings on your monitor will affect other readings you might obtain.

Rule four: Understand the relationship between flammability and toxicity: Understanding the relationship between toxicity and flammability when measuring is very important. Flammable range is a measurement (usually in % of the total volume of air/ignitable vapor mixture as shown on your monitor) of the amount of a substance that will ignite when mixed with air. When a substance is present in sufficient quantity to be measured as an ignitable mixture, it can also be measured in relationship to its total volume in the mixture, represented in Parts Per Million (PPM). PPM is the most frequently used method to measure toxicity, while percentage is the most frequently used method to measure flammability levels.

Rule five: A substance (vapor) that comes after you is much more dangerous than one that expects you to come to it: These types of products require great caution since the product or its vapor (which may be toxic or flammable) does not require you to get "intimate with it" to cause a problem. In other words, the substance is coming to find you; you don't have to go find it.

Rule six: Know your monitor's operational parameters: This includes issues like:

1. How long are the sensors in your monitor accurate for—1 year, 2 years?
2. Is the monitor RF shielded to cut down on potential interference from electrical and radio sources?
3. If using a hand aspirator, how many pumps are necessary for each foot of tubing to bring the product into the sensor housing?
4. Do you use a water filter on the end to prevent liquid from pulling up the tube and subsequently ruining your monitor?

All of these items should be in the technical manual that came with the monitor you use. Remember that you should specify training as part of the purchasing package.

Rule seven: Battery operated monitors will not work if the batteries are dead: Checking the batteries at the incident is a novice approach to problem solving and not very smart. Batteries should be checked regularly (each shift).

Rule eight: Zero and Field calibrate (bump check) your instrument in clean air: Before using any monitoring equipment it should first be checked to ensure it is reading zero (0) for flammability and toxicity and 20.9% for oxygen. Follow the manufacturer's recommendations for your monitor on field calibrations. Make sure you are in clean air when you do this or your readings will be inaccurate.

Rule nine: Sample from upwind: Standing upwind will allow you to approach the potential hazardous atmosphere at your own pace. Never let the wind bring it to you! Good old common sense stuff!

Specific Monitoring Measurements

Oxygen

Monitors usually measure oxygen concentrations between 0% and 25% in air. Your monitor should be set up (since it's designed for use in confined spaces) to alarm at 19.5%, which is the minimum adequate percentage of oxygen concentration established by OSHA. It should also be set to alarm at levels of oxygen above 23.5% in the air.

Normally air consists of about 20.9% oxygen. Oxygen deficient atmospheres are those with 19.5% or less. Oxygen enriched atmospheres are those with concentrations above 23.5%.

The reason you check oxygen at this juncture is because at certain oxygen deficient levels (look at your manufacturer's data) the flammability readings you are about to take are invalid or altered. Also, at levels above 23.5% your measurement of flammability will not be accurate and will render false readings.

Flammable and Combustible readings

CGI sensors in your monitor are known as combustible gas indicators. They determine the presence of flammable vapors of hydrocarbon products that might be present in the trench. There are certain instruments that are designed to measure a single product (for instance, methane) only; they would measure the flammable vapors as a percentage of the lower explosive limits. The monitor that you are using is calibrated for a certain flammable gas, either methane, pentane, butane, or hexane. You should test for flammability during any of the following potential scenarios.

1. Any suspected contaminated trench
2. As a process of leak detection
3. If you are investigating an unknown material

The monitor you are using has a preset alarm level for 10% of the lower explosive limit (LEL). That means that when the level of the product you are testing (actually the level of the product the monitor is calibrated for) reaches 10% of its lower flammable limit the monitor will sound an audible alarm as well as provide a visual signal. Does this mean the world is going to blow up? Not at all. You, in fact, have a 10 to 1 safety factor built into the alarm system. When you "alarm," you are only at 10% of the lower explosive limit. In order for the atmosphere you are in to actually ignite, you must reach 100% of the LEL, still more than 90% away. This is your action level, the level at which you need to make a decision, not the level at which you have to panic!

At a minimum atmospheric monitoring should be done for oxygen, flammability, and toxicity

Remember that we tested oxygen first. The reason is that CGIs must have a certain percentage of oxygen present in order to function properly. Most instruments require a minimum of 10% oxygen to operate, but they may be inaccurate at levels even higher than that. Many must have as much as 16% oxygen in order to give an accurate reading. (Look at your manufacturer's literature and limitations). So, if you assess oxygen first, it provides the information you need to determine if your flammability readings are going to be accurate and usable as a tactical tool.

Measuring toxicity

The monitor you are using has either one or two toxic sensors, most likely set up to measure Hydrogen Sulfide (H2S) or Carbon Monoxide (CO), the two most common toxic vapors. Your action limits (alarm settings) are at 35 ppm for Carbon Monoxide and 10 ppm for Hydrogen Sulfide. Again, does this mean it's time to panic? Absolutely not, these are the Time Weighted Averages for an OSHA 8-hour exposure. This indicates that breathing apparatus must be worn and that a problem exists in the trench, which you are going to attempt to control or eliminate.

Putting it all together

So you're standing at the trench with the monitor in your hand. You managed to read all the manufacturer's literature, and you know exactly how the monitor works. You have nurtured, calibrated, bump tested, and talked nicely to your monitor as you lowered the sampling probe into the trench. The question now is what do you do with the data you are capturing?

Monitoring should take place prior to entry and at least every five minutes during trench operations. This is not a hard-and-fast time frame, and it should be adjusted based on the severity of the atmosphere you are dealing with. Concerns about changing or presently dangerous atmospheres should prompt you to monitor more frequently. Consider the following guidelines when monitoring:

1. One person should collect and record monitor readings throughout the entry and rescue/recovery operation. This should be his sole assignment.

2. All readings should be captured on the trench rescue tactical work sheet (see Appendix 4).

3. Readings should be reported to the Extrication officer or the Operations officer on a continual basis.

4. Any fluctuations or changes in readings should be immediately reported.

5. Any alarm levels should immediately be reported and action taken.

6. Never leave the monitor unattended. It could get kicked into the trench, stepped on, or ignored. It's not a glorious job but someone has to do it.

7. Always use a Haz-Mat Team to your best advantage.

HAZARD CONTROL USING VENTILATION

Ventilation: Ventilation is the first method of choice during rescue operations since it is fast and easily monitored. The point is that you can't expect ventilation to work for you every time. You must use your monitor as a guide and be prepared to make hard decisions based on facts.

If it's an atmospheric problem, it makes sense that you might be able to control it

Ventilation for atmospheric hazard control

with ventilation. Ventilation is only as good as the technique you employ and, of course, the nature of the product. If you have a flammable reading at an action (alarm) level and you begin to ventilate, are the readings dropping to acceptable levels? If not, why? Is it because your ventilation is not working, or is it because there is a liquid product that continues to give off vapors despite your best efforts?

If you are using ventilation as a hazard control method in trench rescue, you will have to consider the outside temperature and the affect ventilation will have on your victim and rescuers. In addition, multiple fans may be needed on intersecting trenches.

Leave it alone: If it is a Haz-Mat call, let the Haz-Mat Team deal with the problem. Once hazards associated with atmosphere or contamination have been eliminated, go to work.

FINAL THOUGHTS

Now that you understand the concepts and applications of atmospheric monitoring, consider enforcing the following concepts with your personnel:

- Read the instruction manual inside and out, be very familiar with it, become the expert with that monitor!

- Practice, practice, practice. Utilization of your monitor(s) must be second nature to you. You should be able to scroll through them with ease, understand their idiosyncrasies, and operate them effectively in all types of conditions.

- If at this point you are completely lost, you would be best served to identify that one guy or girl in your organization that is the atmospheric monitoring guru and who acts like a sponge—usually one of your Haz-Mat Team members or an industrial hygienist.

Summary

Trenches by definition are not confined spaces; however, there are many similarities in the definition of a trench and a confined space. For this reason, and to ensure the health and safety of your personnel, it is prudent to make atmospheric monitoring an important part of your on-scene assessment.

Atmospheric monitoring is used to determine the presence of Immediately Dangerous to Life and Health (IDLH) substances, many of which could kill or disable your rescue workers on the scene of a trench incident.

Considerations for atmospheric monitoring in the trench environment include: nature of the hazard being monitored, electrical interference, environmental site conditions, and the presence of certain gases and vapors, all of which could interfere with your monitor's readings.

Action guidelines are preset alarms in your monitor that sound an alarm and indicate that some action on your part is necessary. The action guidelines are:

1. oxygen – less than 19.5% or greater than 23.5%
2. combustible/flammable gas – 10% of LEL
3. toxicity – CO 35 PPM and H2S 10 PPM

Using ventilation as a hazard control option for IDLH atmospheres should take into consideration the nature of the product, using your monitor readings as a guide to determine successful ventilation, the wind direction, temperature and humidity, and whether the configuration of your trench inhibits ventilation because it intersects.

Questions (Answers and Discussion on page 254)

1. Trenches that are four feet or more in depth must be atmospherically monitored when it can reasonably be expected that an atmospheric problem may exist. This monitoring should be for which of the following:

   a. Oxygen, chlorine, and gasoline
   b. Oxygen, flammability, and toxicity
   c. Flammability, corrosiveness, and sulfur
   d. Any foreign product that may injure a worker

2. When a preset alarm activates in your monitor it indicates:

   a. There is a very dangerous situation at hand
   b. Some action on your part is necessary
   c. Evacuation of the scene should be ordered
   d. All of the above

3. When you are using ventilation for hazard control you should:

   a. Consider the air temperature
   b. Consider the nature of the product being monitored
   c. Consider the humidity
   d. Consider if intersecting parts of the trench are hampering ventilation success
   e. All of the above

4. Your response to a trench incident should include the Haz-Mat team.

   a. True
   b. False

UNIT SIXTEEN

GAINING ACCESS

TERMINAL OBJECTIVE

Determine the steps and considerations in gaining access
to the trapped or buried victim of a trench emergency.

ENABLING OBJECTIVES

The student will be able to:

- Describe the two types of situations presented to a rescuer at a trench collapse

- Explain the methods that can be used to uncover trapped or buried victims

- Specify the rules to follow when digging for a trapped or covered victim

At this point, let's review a case in which a worker was trapped under a pipe in a trench where the contractor's protective system was in place and functional at the time of the incident. Upon arriving at the scene, the victim was found to be conscious but not alert, and therefore was considered to be a true emergency victim. He actually appeared to be in some sort of trauma-related shock.

As a result of the initial victim assessment, a ladder was placed in the trench and a paramedic entered to conduct an initial victim survey. Immediately, the medic became confused and disoriented. In fact, he was never able to reach the victim, and slumped over at the bottom of the ladder. Realizing that something was wrong, the incident commander ordered another firefighter to enter the trench with an SCBA. This firefighter, with full protective clothing and SCBA, climbed down the ladder and retrieved the medic. He also was able to place a victim Supplied Air Breathing Apparatus (SABA) on the victim.

This story obviously deserves a closer critique. As it turns out, the worker was trapped by the pipe but not seriously hurt. He was never in danger of dying from any trauma-related injury because the pipe had only trapped his foot in the soft sand of the trench floor. He had actually fallen victim to the fumes from the solvent can that he was using to clean the pipe. The incident commander assumed the trench was safe because the contractor's protective system was in place. The victim's confusion and disorientation was taken as a sign of shock, when he was actually overcome by vapors.

EXERCISE: Using the assessment skills you have learned, would you have properly handled this emergency?

This serves as an example of the problems that may confront an incident commander at the scene of a trench incident without a collapse. Other than having live or dead victims, there are really only two types of incidents that can occur at a trench site: those incidents that involve a cave-in, and those that do not involve a cave-in. It is imperative that each of these situations be evaluated with the same considerations, regardless of the circumstances. Following the same type of evaluation criteria for both types of incidents will keep you from committing your personnel to an unsafe area or a potentially hazardous environment.

Trench accidents that do not involve a cave-in will almost always be more difficult to handle than those in which a cave-in has buried your victim. This is because it is more stressful to deal with a situation when someone's life is hanging in the balance. If the person is dead, it doesn't matter what you do. It won't affect the outcome for the victim (but remember to protect your personnel).

To say that your skills will be tested would be an understatement. Whether it is a pipe that has broken a rigging strap and fallen, equipment failure, toxic environment, or victim illness, your success will be determined by following a safe and logical plan. A plan that does not skip steps!

Accidents with a cave-in fall into two different categories. Those involving a partially buried victim, and those in which the victim is completely buried. Both types will be challenging, since each may involve a substantial amount of work, depending on the entrapping mechanism.

The types of material that can entrap a worker are as numerous as the types of equipment found at the construction scene. Pipes and other heavy objects will have to be lifted off the victim, using airbags or another type of mechanical lifting device. These lifts can be done from inside the trench or from above the trench using lifting straps.

> **TIP:** The cardinal rule for gaining access to the victim is to remove any entrapment mechanism and uncover the victim's head and chest first.

Buckets can be used to remove dirt from the trench and lower tools to rescue personnel

If dirt or sand is the entrapping mechanism, then get out your entrenching tools and go to work. You're in for a long couple of hours, or even days. The magnitude of this type of situation is never more evident than the first time you have to use a rope to raise a five-gallon bucket half filled with dirt. In many cases, tons of dirt may have to be removed from the trench by hand. Another good job for a "Mongo."

If you are lucky enough to live in a more populated area, there is a good possibility that one of your local utility companies has a soil vac truck. These trucks are designed to suck up dirt and other small debris and deposit it into a tank similar to a septic tank vac truck. This is another of those items that can be identified and preplanned during your preparation assessment.

Excavator vac truck

When digging in a trench to remove a buried victim, there are a few rules you should follow regardless of the victim's condition. Rule one is to never use a mechanical device or backhoe to dig up or pull out a partially buried victim, that is unless you subscribe to the old saying "parts is parts." Rule two is to never attempt to pull a partially buried victim out. The reason is: refer to rule one on parts. Rule three is to dig by hand when you get near the victim. Shovel marks on the victim's face are hard to explain to the coroner.

> **TIP:** Remove all dirt from the trench when you are digging to remove a victim. Resist the temptation of just shoving it to one side because, guaranteed, you will be moving it again.

Summary

There are two types of incidents that can occur as a trench emergency: those in which a collapse has occurred and those in which an injury or entrapment has occurred with no collapse. Accidents that involve a collapse either contain a completely buried or partially buried victim.

There are many types of potential entrapment mechanisms for workers in a trench. Equipment failure, broken rigging straps, and operator failure are just a few. Remember the cardinal rule for gaining access to a victim is to remove any entrapment mechanism and uncover the head and chest first.

If dirt is the entrapment problem, then depending on the nature and the amount of soil, you could be in for a work intensive call. Much of the soil may have to be removed by hand using small entrenching tools and buckets. Having the availability of a vac truck to remove loose soil can be a big benefit in these types of collapses.

When digging to remove a buried victim, never use mechanical equipment or attempt to pull a partially buried victim out with a mechanical device. Always dig by hand when around the victim and, as a rule, remember to remove all dirt from the trench as you dig.

Questions (Answers and Discussion on page 255)

1. Which of the following methods concerning digging operations is unacceptable:

   a. Digging with entrenching tools
   b. Using a backhoe to pull away material that is remote from the victim
   c. Digging by hand
   d. Using large spoons and juice containers

2. When gaining access to a victim of a trench collapse the primary concern is to:

   a. Evaluate the airway and provide oxygen
   b. Uncover the head and chest first
   c. Assess any major bleeds
   d. Provide warm IV fluids

3. Accidents that involve a cave-in fall into two different categories:

   a. Those in which the victim is either alive or dead
   b. Those in which the victim is buried or partially buried
   c. Those that contain a disabled and non-disabled victim
   d. Those in which the victim is conscious or unconscious

UNIT SEVENTEEN

PROTECTIVE SYSTEMS IN TRENCH OPERATIONS

TERMINAL OBJECTIVE

To understand the proper use of sheeting and shoring for normal and specialized trench rescue operations and the traditional methods used by contractors for safe trench and excavation openings.

ENABLING OBJECTIVES

The student will be able to:

- Explain the proper use of sheeting for trench rescue operations

- Describe the proper use of shores for trench protective systems

- Describe techniques for using isolation tunneling for victims trapped in running debris

- Explain the use of shaft tunnels to reach buried victims from remote locations

- Specify the procedure for building a Class C protective system

- Describe the various methods contractors use to stabilize trenches and excavations

- Explain the different components and materials used by contractors to shore a trench

SHEETING AND SHORING FOR TRENCH OPERATIONS

SHEETING

Of all the techniques that can be used to make a trench safe, by far the one you as a rescuer will use the most is sheeting and shoring. This is for a number of reasons, but mostly because commercial sheeting systems are designed to be installed before digging begins, or they require near vertical walls to be effective. Trench collapses don't normally come with all the walls straight; if they did, you wouldn't be there in the first place.

The panels used in the rescue business contain the sheeting with strongback attached, and can be lowered into the trench at just about any angle that will allow you to get a shore shot from one strongback to the other. This being the case, a trench training class produces much nicer trenches than a trench collapse call. For example, on real events you will find that panels and shores often do not line up. Don't worry about "pretty." Concentrate instead on the physical forces involved in the collapse and what you are trying to accomplish.

Setting panels in a trench is accomplished using a variety of methods depending on the condition of the trench and location of the victim. In most cases, the strongback will be facing the inside of the trench and the panel will be as close to the wall as possible. Care should always be taken when setting panels that one or more of your "Mongos" don't end up in the trench! You know how it is when you get three or four of them together.

When setting panels, you will have the option of doing a "same side" set, or an "opposite side" set. If you can do a same side set, the panel is moved into place and the ropes that are tied on the bottom of the panel are held, while the top of the panel is pushed out and over the trench. The ropes are then used to slowly lower the panel, keeping it tight to the wall. If adjustments of the panel are necessary after setting, the ropes are just pulled up and the

Same side panel set in T Trench

panel is moved. As a rule, same side sets are the easiest and fastest, but require more room to maneuver your panel team. This ultimately could involve the moving of a substantial amount of the spoil pile.

TIP: As a rule of thumb, you should strive to create a safe area of 12 feet in the trench.

Opposite side panel set using 2 x 4 runners as guides

When it is not possible to set the panel from the same side of the trench, you can use the opposite side technique to set the panels using two 4" x 4" runners. The "runners" are placed against the opposite side wall with one end of each runner touching the bottom of the trench on the opposite side, and the other ends on the top of the side you are on. The panel is then flipped so that the strongback is pointing down. This will allow you to slide the panel down the runners until the bottom of the strongback and panel are against the opposite wall. You can then push the panel upright to the other side of the trench using a pike pole.

TIP: The opposite side set is also a good set when you have a victim in the trench and additional care has to be taken with panel placement.

While there may be some variation in how the panels are set, there is no variation in how many you set. For non-intersecting trenches the minimum number of panels that will be needed is six. The first two panels are set on either side of the victim and then two more on each side. This creates a working area for the rescue team if digging or care is needed. In all cases, you should give consideration to setting the first two panels on each side of your victim.

TIP: Secondary collapse potential is a very real possibility, especially after dirt removal. Make sure your victim is protected by setting panels on each side of the victim first.

SHORING

Once the panels are set, you will want to make the system safe by installing the shores. Shores are the component of the system that will transfer the forces across the trench, up the strongback and then into the opposite sidewall. This will complete the protective system and give you a safe area in which to work.

Pneumatic Shores

Commonly referred to as "shoot-ing the shore," the air system is con-nected to the shore and it is lowered into place using ropes attached to ei-ther end. When the shore is lined up in the proper place (usually done best by someone at the end of the trench), the signal to "shoot" the shore is given. The shooter will then activate

Rescuer shooting and setting a shore

the air, which allows anywhere from 200 to 250 psi of pressure to extend the

Paratech Duel Deadman Controller used for shooting the shore

shore. At this point the individual shore type will determine the method that you will use to lock it off. Some air shores are manually pinned and some lock automatically. In any case, after securing it manually, the pressure is released and the air hose removed and placed on another shore. Keep in mind that the manual systems used to tighten and lock shores can exert far greater pressure than the air used to shoot the shore. In some cases this pressure can exceed 450 psi.

TIP: The only person who can call for the shore to be shot is the installer. This will help keep him from getting hurt by the shore being accidentally activated.

Once the shores are in place they create an overlapping zone of pressure on the trench wall. This overlapping pressure stabilizes the trench walls and prevents collapse.

The placement of shores in the trench is also completed in a predetermined order. When dealing with air operated shores you will want to shoot the middle shore first. If you think back to the physical forces discussion, you will remem-ber that the most unstable part of the trench is about three-quarters of the way down from the top. For this reason you will want to install the middle shore first.

After installing the middle shore you will move your efforts to installing the bottom shore. This is the next most dangerous and unstable area. Following the same procedure as with the middle shore, the rescuer will climb down a ladder beside the panel and shoot the shore. Care should always be taken not to allow the rescuer to get further into the trench than waist high to an installed shore. As an example, the person shooting the bottom shore would be allowed to enter into the trench between the panels until his waist was even with the previously installed middle shore. It is easy then to lean over and secure the shore according to the manufacturer's recommendations.

> **TIP:** Always have the first shoring team members in a Class 3 harness with tag line while placing the first set of panels and shores.

The top shore would follow, with care being taken not to shoot it too close to the trench lip. The rule of thumb for shoring is to install the middle shore, and then split the difference between the middle and top and the middle and bottom for the other shores. It is this author's recommendation that three shores be shot for all trenches six feet deep or more, regardless of the type of soil. Regardless of how many vertical shores you use, the maximum distance between shores horizontally should never exceed four feet.

> **TIP**: Remember to always toe nail the end of pneumatic shores to the strongbacks. This way, if subsequent shores cause the system to become loose, previously installed shores won't fall out.

Always toenail struts so they don't fall out if tension is released

Timber shores utilize a somewhat different procedure for installation. Because a timber shore needs to be measured, cut, and scabbed after placement, the top shore is installed first. Subsequently, the middle shore is done and then the bottom. With each installation, the rescuer is only allowed to enter the trench at waist level to the last secured shore.

Scabbing around a 4 x 6 timber shore

Because cutting the timbers for shoring is difficult at best, scabs are used to hold the shores in place. It is usually easier to nail the bottom part of both scabs to the strongbacks and use this as a shelf on which to rest the shore while installing the top scab. Wedges can then be used to tighten the system. Keep in mind that the minimum size Douglas fir shore that is approved for Class A and B trenches up to 10 feet in depth is 4 x 6.

TIP: 2 x 4 rails can be installed on the strongback previous to setting panels. This will make shore adjustment easier.

Wales are horizontal members used to span openings along the trench walls, and they can be of the inside or outside variety. As we discussed earlier, they can be made of timber, metal, or makeshift ladders. The length of the wale is entirely dependent on the area that needs to be spanned.

2 x 4 runners and scabs can be preplaced on panel prior to set to make shore installation easier

Inside wales are used to span a trench panel for the purpose of creating a safe area in a "T" Trench, or to create an open space in the bottom of a trench. The wales are lowered into the trench on the inside of the panels, and placed up against the strongback. The shores are then shot to the wales.

Airshores with heavy duty rails as wales

Outside wales are placed against the trench wall before placement of the panels. They are used to span an opening that may have been created by a slough of the trench wall. After installation of the wales and panels, but before the shores are shot, it is necessary to backfill the area between the wale panels and the trench wall. This will allow the forces to effectively be transferred when the protective system is complete.

Inside wales used to create an open space for victim access

TIP: All areas of the trench need to be backfilled when there is an open space between the protective system and the trench wall. If a void is not present, no transfer of forces can take place.

Outside wales and airbag backfill
prior to panel set

Supplemental sheeting and shoring is necessary when an area of dirt is removed or falls from behind the panel and into the trench. This creates a void that will need to be filled if it extends more than two feet below the bottom of the protective system. To address this problem you would construct and install supplemental sheeting and shoring. This type of shoring is nothing more than additional shoring put together and installed to address the void area.

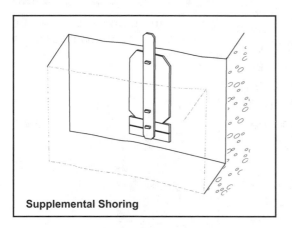

Supplemental Shoring

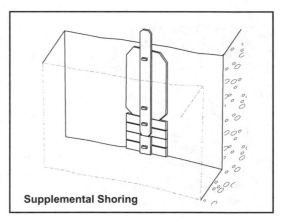

Supplemental Shoring

TIP: If you are concerned with the stability of your protective system, you can always add additional strongbacks or wales, shorten your shoring zone, or call an engineer for advice.

ISOLATION TUNNELS, SHAFTS, AND ENGINEERED CLASS C SYSTEMS

There is always the possibility that the soil profile will not allow you to excavate an area without the product running back in the trench as fast as you can remove it. Sand, coal, grain, and other similar products are running debris and therefore hard to sheet and shore. In these cases, special systems called isolation tunnels and shafts can be built.

Isolation tunnels are cylinder objects that are placed over top of your victim and then worked around them as the material is moved. In the case of a coal worker who was trapped in running debris, a 55-gallon barrel with both ends cut out was used. Keep in mind that objects like the barrel are strong in the vertical orientation, but weak if laid horizontally. Concrete pipes can also work for this method.

Isolation vessel used for worker trapped in running coal

A similar method can be used to build and install a shaft. In the case of the shaft, the system will take longer to construct than other types of systems, but this may be your only option for a worker who is trapped in running debris.

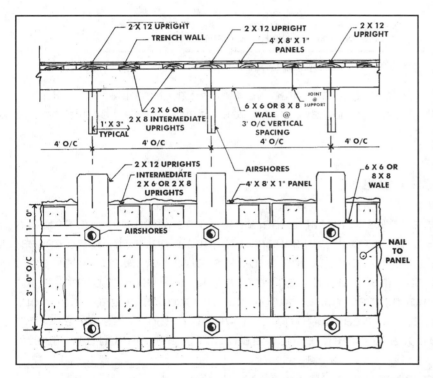

The Engineered Class C System is a special system designed by an engineer to address problems created by the very worst Type C soil. In this type of system, each 4' X 8' piece of sheeting panel has three strongbacks. These strongbacks are toe nailed to the sheeting and then held in place using 6" x 6" wales. The shores are then shot to the point where the wale crosses the center strongback of each panel. This type of system utilizes many contact points in order to transfer the tremendous forces created by a Type C soil surcharge.

EXERCISE: Refer to the Class C illustration and count the number of contact points that are created. How many additional contact points are created when compared to a normal trench with wales?

**PNEUMATIC SHORES
WALER SYSTEM
SOIL – TYPE C
Pa = 80 X 72 psf (2" SURCHARGE)**

DEPTH OF TRENCH (FEET)	AIRSHORE SYSTEM (4' X 8' X 1" PANEL WITH 2 X 12 UPRIGHTS) *		WALES		UPRIGHT INTERMEDIATE TIMBERS NAIL TO PANEL **
	MAXIMUM HORIZONTAL SPACING (FEET)	MAXIMUM VERTICAL SPACING (FEET)	NOMINAL SIZE (IN.)	VERTICAL SPACING (FEET)	
UP TO 8 FEET	4' O/C	3' O/C	6X6	3' O/C	2-2 X 6
OVER 8 FEET TO 16 FEET	4' O/C	3' O/C	8X8	3' O/C 2-2 X 8	

* PANEL & 2 X 12 OF ARCTIC WHITE BIRCH
** 2 X 6 AND TIMBERS SYP NO. 2 MINIMUM

Example: Engineer-approved tabulated data for airshore

In all cases, in order to create a protective system that is different from the OSHA standard, it must be certified and approved by a registered engineer and maintained on the scene.

COMMERCIAL TECHNIQUES

As much as we may hate to admit it, there are times when our equipment's capability to address a trench collapse situation is questionable. In these cases, professional engineers and construction people should be called in to assist you with the extrication effort. This doesn't mean you give up responsibility for the situation. It just means that you recognize the situation is greater than your abilities as a rescue team. Keep in mind that we don't dig holes for a living; construction people do, and therefore they are much better at it than we are.

Consider using commercial techniques and professional help if any of the following occur:

- The trench is deeper than 15 feet
- There has been a massive cave-in
- Workers are trapped in running debris
- Environmental conditions prohibit the rescue effort

Typical contractors trench box

Any protective system that is put in place for either rescue or commercial operations must be capable of withstanding all intended and reasonably expected loads. Protective systems are "engineered" to withstand these forces. The decision regarding the type of protective system to use may vary based on the type of rescue operation or, in the case of a commercial situation, the type of job, location of adjacent structures, and time.

Regardless of the parameters, protective systems are based on a set of factors that are evaluated during each operation. These factors may include but are not limited to the following:

* Adjacent structures
* Existing hazards
* Soil type
* Water profile and hydraulic table
* Depth and width of the trench
* Purpose of operation (rescue vs. utility installations)

SLOPING OR BENCHING SYSTEMS

Sloping or benching is used as a method to decrease the angle of a wall to a point that it does not want to collapse. This method reduces the gravitation forces and total amount of unconfined compressive force that is present. It may also be applicable during a rescue operation or during a body recovery as a method to "safe" a large opening. Consider the following when evaluating the need for sloping or benching:

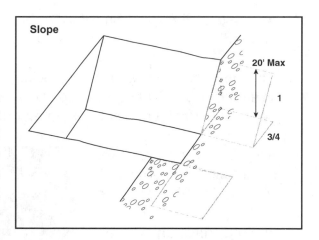

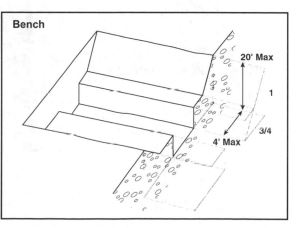

* Requires tabulated data
* Requires the use of heavy equipment
* Is time consuming
* Takes a lot of room

TIP: Sloping may, in some cases, be referred to as cutting back to the angle of repose. This is the point where the material can support its own weight and is not expected to flow. Keep in mind that any sloping at the scene of a trench collapse should be at least 1.5-foot horizontal to 1-foot vertical.

SUPPORT, SHIELD, AND OTHER SYSTEMS

Other options that you may consider for commercial protective systems are support and shield systems. These may be rabbit boxes, coffins, or other trench boxes that are constructed of wood, steel, or aluminum, and are commercially available. In general, keep the following in mind when looking at commercial systems or designing a system to use during a rescue operation:

Aluminum trench box

- Any use of timber should follow the shoring appendices in the OSHA standard.

- Shields, struts, screw jack, beli-jacks, hydraulic speed shores or air deployed shores should be used in designs based on the manufacturer's tabulated data.

- Any deviation from the manufacturer's data needs to be in writing and on site.

- You may have an engineer design a system for your specific environmental needs, but copies of the data need to be on site.

Interlocking sheet panels

Summary

Sheeting and shoring are the primary tools used to create a protective system in a trench collapse operation. Sheeting panels have a strongback connected to them that is used to place the ends of the shores. In most cases the strongback is the only part of the protective system considered when evaluating the trench for OSHA compliance.

There are a number of techniques used to place panels in the trench. Two of the most effective are the same side and opposite side panel set. The same side set is faster and more efficient while the opposite side can be used when you cannot access one side of the trench or you have a victim near the panel set site. Always set your first set of panels on either side of the victim.

There are a number of different shores and shoring systems used to construct protective systems for trench operations. By far the most popular in the rescue business are pneumatic and timber shores. Shores have a proper order in which they are installed depending on the type of shore being used. Supplemental shoring is necessary when digging operations require the removal of soil and it ends up two feet below the bottom of the strongback.

Horizontal members that cross the strongbacks in a trench are called wales. Wales are used to span an opening in the trench wall or inside to create an open space in the trench.

In some cases the magnitude of the collapse prohibits the use of normal rescue techniques. In these events, commercial techniques are necessary. Deep trenches, those over fifteen feet, and/or with running debris are a few examples. In these types of events isolation tunnels, shafts, trench boxes, and sloping/benching may be necessary to resolve the incident.

Questions (Answers and Discussion on page 256)

Note: Some of these questions will require you to use knowledge gained in other units of this manual. The bottom line is that this is where you have to start putting all of the tools together!

1. Traditional sheeting and shoring operations used by the rescue service are typically only good in trenches up to:

   a. 25 feet
   b. 15 feet
   c. 30 feet
   d. 10 feet

2. The first set of panels placed at the scene of a collapse should, if at all possible, be placed:

   a. At the weakest part of the trench
   b. On either side of the victim
   c. The area that is easiest to access
   d. Each operation is different and therefore you cannot anticipate where panels are set

3. There has been a massive collapse and you have no walls left that you feel are adequate to support a panel and shore protective system. You may consider:

   a. Using a crane to place a trench box for protection
   b. Sloping or cutting back to the angle of repose
   c. Working without the protective system because the chance of a secondary collapse is past
   d. Both A and B are correct

4. When sheeting and shoring a trench we are making an attempt to stabilize the unstable and keep the stable from becoming unstable. We accomplish this by:

   a. Taking away the potential for the soil to become active
   b. Redistribution of the forces within the trench wall
   c. Taking away the effects of unconfined compressive force
   d. All of the above

5. Inside waler systems are used for a number of reasons. Considerations in the use of inside wales would be to:

    a. Reduce the amount of shores required to build the protective system
    b. Create an open space
    c. Create more upright contact points with fewer shores
    d. All of the above

6. If the stability of your protective system is in question, consider which of the following options:

    a. Add additional strongbacks to each panel
    b. Add inside wales to the system
    c. Shorten your shoring zones
    d. Call an engineer for an evaluation
    e. All of the above

7. The procedure for installing pneumatic shores is:

    a. Top, middle, bottom
    b. Bottom, top, middle
    c. Middle, bottom, top
    d. None of the above is correct

8. The procedure for installing the initial set of timber shores is:

    a. Top, middle, bottom
    b. Bottom, top, middle
    c. Middle, bottom, top
    d. None of the above is correct

9. When digging operations involve the removal of more than _____ feet of soil below the strongback or panel, you are required to provide supplemental sheeting and shoring protection.

    a. 4
    b. 3
    c. 2
    d. 1

10. The opposite side panel set is used to:

    a. Set panels on the same side of the trench as the victim
    b. Protect the victim from being hit by a panel that is being set
    c. Set a panel from the opposite side if one side of the trench is not acces-
       sible
    d. Both b and c are correct
    e. All of the above are correct

11. During shoring operations, the shooting of the shore is accomplished by a
    command given:

    a. By the extrication officer
    b. By the air supply officer
    c. By the shore installer
    d. By the safety officer

12. When placing panels at any trench collapse you should always strive to
    create a safe working area by providing a minimum of _____ feet of pro-
    tected area:

    a. 18 feet
    b. 12 feet
    c. 20 feet
    d. 10 feet

13. The maximum distance between horizontal shores should not exceed:

    a. 5 feet
    b. 4 feet
    c. 8 feet

14. Which of the following is not true about isolation tunnels:

    a. Pre engineered structures should be used
    b. An isolation tunnel is used to isolate a victim from a given environment
    c. 55 gallon-welded steel drum can be used horizontally
    d. Placement of these devices will most likely require rigging and heavy
       equipment

15. Sloping and benching is a protective system method approved by OSHA to provide safety for workers in a trench.

    a. True
    b. False

16. There is a variety of shoring systems available on the market today. When using a pneumatic shore that has a T-handle device or Acme thread, the user can exert _____ additional levels of pressure on the wall by manually turning and securing this device.

    a. Not more than100 psi
    b. Not more than 200 psi
    c. Up to and in some instances more than 450 psi
    d. Never more than what can be gained by the pneumatic pressure of the shore.

17. The first shoring team member to enter the trench for placement of shores should be in:

    a. Class 2 harness
    b. Class 3 harness
    c. Class 3 harness with tag line
    d. Tag line tied around his waist

18. Commercial techniques should be considered if:

    a. The trench is over 15 feet deep
    b. There has been a massive cave-in
    c. Workers are trapped in running debris
    d. Environmental conditions prohibit the rescue effort
    e. All of the above are correct

UNIT EIGHTEEN

VICTIM PACKAGING AND
TERMINATION PROCEDURES

TERMINAL OBJECTIVES

To understand the methods, techniques, and equipment used
for trench rescue victim packaging.

To understand the elements to be considered as part of the
termination process for trench rescue operations.

ENABLING OBJECTIVES

The student will be able to:

- Explain the considerations for trench rescue victim packaging

- Describe the techniques for victim removal

- Specify the various victim packaging equipment utilized in trench rescue operations

- Explain why the termination process can be the most dangerous phase of the operation

- Specify the order in which the trench is dismantled

- Describe the importance of proper clean-up procedures after a trench rescue operation

- Summarize the conditions that may lead to critical incident stress debriefing for your personnel

VICTIM PACKAGING AND REMOVAL

Trench rescue victim packaging and re- moval techniques are not all that different from other techniques you have been previously taught (with the exception that you are going to get dirty). There are, however, a few con- siderations that warrant your attention.

A ladder can be used as an incline plane to assist in victim removal from the trench

While working to remove a victim from a trench, extreme care must be taken not to dis- lodge any of the shoring material. This can, and frequently does, happen during extended digging operations when many personnel are working at the same time. Someone on the scene should con- tinuously monitor the integrity of the protective system.

Essential to your successful removal of a buried victim is careful and comprehensive preplanning of activities leading up to the re- moval. Time and again, rescuers place their shoring material so close to the victim that it is impossible to remove them. Take the time to forecast your movement patterns, and how the type of packaging device will affect the victim's removal. If planning reveals that ad- justment, or moving of the shores is neces- sary, start before victim packaging begins.

Victim packaging in a trench is often in tight quarters. Preplanning for victim removal is a must.

Another concern of victim extrication from a trench is something that is usually handled by your rigging sector. The victim in the hole will eventually have to be lifted out of the trench. If the victim is large, or the packaging device is cumbersome, you may want to retrieve them with some sort of mechanical advantage system. To effectively do this, you will need some kind of elevated point of attachment.

Ladder derrick to provide over- head lifting point

TIP: I recommend that you review Michael G. Brown's book *Engineering Rope Rescue Systems* for a comprehensive review of mechanical advan- tage systems.

TERMINATION PROCEDURES

The most dangerous phase of your rescue operation will be the termination. It is at this point in the process that the adrenaline is gone and the personnel are tired. Special care needs to be taken to see that rescuers are rested and alert before breakdown begins. You may well have to call in fresh crews to facilitate the termination and breakdown of the equipment. At any rate, don't rush. Give your personnel frequent breaks. Keep in mind that the emergency is over.

As a rule of thumb, the termination and breakdown of the protective system happens in the reverse order that it was built. This process begins with the manner in which the shores are broken down and ends with the removal of the fire line tape that was first put up to control access. This is the part of the event where you really have to keep an eye on your "Mongos." They have a tendency to work fast at all costs. Keep them focused on the fact that the job is not successfully completed until it is terminated and no personnel have been hurt.

Another area that needs to be mentioned is clean up. "Holy - Moly!" You will be confronted with the dirtiest, nastiest bunch of equipment you have ever seen. In addition, it all has to be spotlessly clean before the first guy gets a chance to sip a "brewski." That's the rule! The moral to this story should be that nobody goes home until all the equipment is cleaned and packed.

The last part of your operation will involve a critique and possibly provide Critical Incident Stress Debriefing (CISD). The critique is always important because the event gives you an opportunity to evaluate your performance and make adjustments so that the service delivery system gets better all the time.

For some people, CISD is the part of the event that allows them to defuse and gear up for the next one. Just remember that the way something affects you is not always the way it affects someone else. Everyone's tolerance for critical events is different. If you have had a significant event, offer the CISD to all who participated in the rescue effort.

Summary

Considerations for victim packaging at a trench rescue should begin before the protective system is designed and installed. This will make sure that shores and other material are not in the way of the victim access route. It will be necessary for someone to continually monitor the trench protective system while victim packaging is underway and additionally to make sure during removal that no shores are displaced. The best packaging device is the one that will allow removal of the victim from the trench without further aggravation of existing injuries.

Termination procedures can be the most dangerous part of the rescue. Don't let your folks get careless because the adrenaline is gone and they are tired. Rehab your folks often and, if necessary, bring in fresh personnel to perform the breakdown.

The incident is never over until the critique and evaluation of the team's performance is complete. Take time to praise your team's efforts but also have them explore ways they could have completed the rescue safer, quicker, and more efficient. Remember to also provide your team members with Critical Incident Stress Debriefing if the incident proves to be unusually hard psychologically for your personnel.

Questions (Answers and Discussion on page 261)

1. Considerations for victim packaging should begin when the victim is uncovered.

    a. True
    b. False

2. During the removal of a victim from a trench collapse your greatest concern(s) should be:

    a. Not aggravating the victim's existing injuries
    b. Choosing a packaging device appropriate for the space available
    c. Getting the victim medical care as soon and as safe as practical
    d. All of the above

3. If a lifting operation were necessary on a trench accident it would most appropriately be handled by:

    a. Logistics
    b. Extrication
    c. Panel Team
    d. Rigging Team

4. Termination can be the most hazardous time of the incident.

    a. True
    b. False

UNIT NINETEEN

TECHNIQUES FOR TRENCH PROTECTION

TERMINAL OBJECTIVE

To understand the proper procedures to build protective systems
for most types of trench collapse emergencies.

ENABLING OBJECTIVES

The student will be able to:

- List the considerations that apply to trench rescues

- Explain the procedures used for constructing a protective system in Straight Wall, Single Wall Slough, "T," and "L" trenches

CONSIDERATIONS FOR ALL TRENCHES

There are certain steps that will be taken on all trench emergencies. For the sake of organization and explanation they are listed below as generic to trenches of all types. That is to say, you will need to consider them on all trench rescues, regardless of the type or technique used for protection. Therefore, after this introduction they will not be mentioned under the specific techniques that follow.

- Provide hazard control to eliminate any existing or potential hazards that could jeopardize the rescue attempt before the protective system is built.

- Always establish an Incident Management System that assigns accountability and responsibility for specific job functions to specific individuals. This will help organize your scene and provide for the necessary accountability of personnel.

Hazard control should include marking all utilities, including underground gas lines

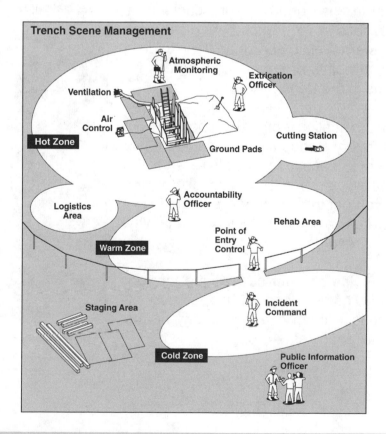

Trench Scene Management

Atmospheric Monitoring

Extrication Officer

Ventilation

Air Control

Hot Zone

Cutting Station

Ground Pads

Logistics Area

Accountability Officer

Rehab Area

Point of Entry Control

Warm Zone

Staging Area

Incident Command

Cold Zone

Public Information Officer

- Monitor the area around and in the trench before and during the extrication effort. Remember to monitor for oxygen, flammability, and toxicity before taking any actions. Refer to the chapter on atmospheric monitoring for action levels and parameters for entry.

Ground pad placement

- Ground pads will be needed on all trench calls to help distribute the additional load produced when rescuers are working in and around the trench.

- Ventilation will need to be considered on all trench collapses, even if not used. Remember that ventilation has advantages and disadvantages. If it is not needed, don't do it!

Ventilation using fire service

- Ladder access should be within 25 feet of all rescuers in the trench. In all cases, ladder placement needs to be one of the first things done, just in case one of the rescuers or a bystander falls in the trench before the protective system is built. The smart thing to do is provide two points of egress and ingress for all persons in the trench.

- Event documentation is critically important from a cost recovery standpoint; that is if there is a mechanism in place to recoup all or part of the cost of the rescue effort. From a legal prospective you can bet that someone is going to get sued if there is a significant injury or a fatality. Keep good records.

- Always provide an after-call critique for your personnel. These calls don't happen very often, so critique them and handle any performance discrepancies.

Ladder access is an essential first step in providing escape route from trench

TECHNIQUES FOR TRENCH PROTECTION

STRAIGHT WALL TRENCH
Timber - Air - Hydraulic

DESCRIPTION OF TRENCH

The straight wall trench will require the rescuer to set a minimum of three sets of panels. One set directly beside the victim and one set on either side of the initial set to provide a safe area for rescuers to work. The number of shores that will be needed per set of panels is based on the soil classification and depth of the trench.

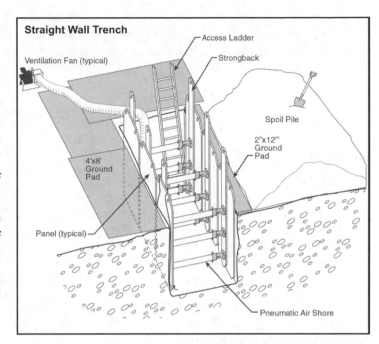

PROCEDURES

- Place middle set of panels as directly beside the victim as possible

- For timber:
    Set top shore
    Set middle shore
    Set bottom shore

- For air:
    Set middle shore
    Set bottom shore
    Set top shore

- For hydraulic:
    Set and expand shores between uprights

- For screw jacks:
    Follow timber guidelines

Straight wall trench using
Airshore Multi shore system

- Set outside panels and shores using procedure appropriate for the type of shore available.

> **TIP:** Rescuers can set the outside panels from inside the safe area of the trench. In that case all shores would be placed middle, bottom, and top.

SINGLE WALL SLOUGH

DESCRIPTION OF TRENCH

This type of trench has a collapse of one wall. In this situation the protective system is designed with outside wales to span the opening and provide a backing for protective panels.

PROCEDURES

- Place pickets to tie wales

- Place and tie off bottom wale

- Place and tie off top wale

- Set middle set of panels as directly beside the victim as possible

- Fill void behind slough side middle panel with air bags or other material

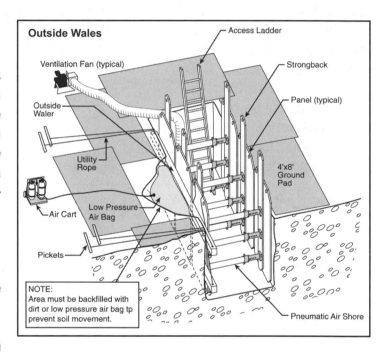

Outside Wales

Ventilation Fan (typical)

Outside Waler

Utility Rope

Air Cart

Low Pressure Air Bag

Pickets

Access Ladder

Strongback

Panel (typical)

4'x8' Ground Pad

Pneumatic Air Shore

NOTE:
Area must be backfilled with dirt or low pressure air bag tp prevent soil movement.

Single wall slough

- Expand to fill void but do not push out panel

- For timber:
  - Set top shore
  - Set middle shore
  - Set bottom shore

For air:
Set middle shore
Set bottom shore
Set top shore

- For hydraulic:
  - Set and expand between uprights

- For screw jacks:
  - Follow timber guidelines

- Expand air bag to tighten system and completely fill void

- Set outside panels

- Fill voids in slough side outer panels if necessary

- Set shores using procedure appropriate for the type of shore available.

6 x 6 timbers used as outside wales

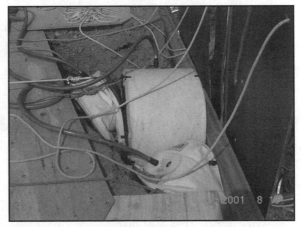

Low-pressure air bags used to backfill slough

TIP: Rescuers can set the outside panels from inside the safe area of the trench. In that case all shores would be placed middle, bottom, and top.

INSIDE WALES

DESCRIPTION OF TRENCH

Inside wales are used in a trench to span a set of panels for the purpose of creating an open space. The open space may be required as the result of a piece of equipment in the trench that cannot be moved, or to create space for a digging and extrication operation.

Inside Wales

Ventilation Fan (typical)
Access Ladder
Inside Waler
4x4 Cribbing
Spoil Pile
Strongback
Utility Rope
4'x8' Ground Pad
2"x12" Ground Pad
Panel (typical)
Pickets
Pneumatic Air Shore

PROCEDURES

- Place bottom wales in trench

- Place all panels

- Shoot middle shore on outside panels

- Raise bottom wale on both sides and tie off to pickets or outside uprights

- Shoot shores between bottom wales

Inside wales used to create an open space by spanning panels

- Place top wales and tie off to pickets or outside uprights

- Shoot shores between top wales

TIP: Additional strongbacks can be added to the system at any time before the wales are shot. The number of strongbacks needed is determined by the soil classification and depth of the trench.

"T" TRENCH

DESCRIPTION OF TRENCH

The intersecting "T" Trench is a very unstable trench, not only because one wall is exposed, but also a section has been cut that intersects the other wall. The key here is to capture the corners as quickly as possible, since they are the most unstable of the exposed areas. Inside wales are necessary to span the center panel because there is nothing to shore against where the "T" leg intersects with the long wall.

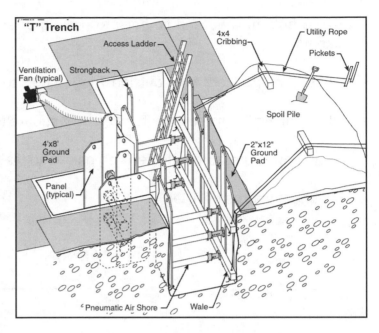

PROCEDURES

•   Limit activity at corners of intersection

•   Set pickets for tie backs (panels and wales)

•   Prepare panels (7) and wales (2)

•   Prepare shores

•   Set two panels on wall of T leg

T-Trench

•   Shoot middle, bottom, and top shore of T leg to initially capture corners (low pressure)

•   Place a bottom wale on trench floor along long wall

•   Set remaining 5 panels
    Two panels on opposite T leg corners
    Three panels on long wall

- Set middle shore on outside T leg and long wall panels (full pressure)

- Raise bottom wale and tie back to pickets or secure to the top of outside panels

- Shoot bottom shore from outside T leg panels to wale

- Place top wale and tie back to pickets or secure to the top of outside panels

- Shoot top shore from outside T leg panels to wale

- Re-shoot shores on T leg

"L" TRENCH

DESCRIPTION OF TRENCH

The "L" Trench can be described as two trenches that intersect at their ends and form a right angle. This type of trench presents a difficult scenario for rescuers because the inside and outside corners of the "L" are difficult to capture with standard protective equipment.

"L" Trench

Labels: Pickets, Strongback, Ventilation Fan (typical), Access Ladder, Panel (typical), Wale, Utility Rope, 4'x8' Ground Pad, Corner Block, Thrust Block, Pneumatic Air Shore

PROCEDURES

- Limit activity at corners of intersection

- Measure depth and width of trench

- Set pickets for tie backs (panels and wales)

- Prepare panels, wales, and shores

- Set opposing inside "L" panels, tie back to pickets

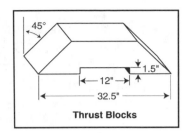

Thrust Blocks

- Place bottom wales on trench floor, both sides, tie back to pickets

- Place "thrust blocks" (one per shore) using joist hangers on inside "L" panels

- Shoot center shores at 50 to 75 PSI to capture corners. Personnel on tag line may accomplish this from a ladder.

- Place and picket top and bottom "kick plates" or 4" x 4" or 6" x 6" timber at bottom edge (toe) and top edge (lip) of outside "L" panels. Extend below lip far enough to picket in place without danger of secondary collapse

Secondary corner brace on L Trench

- Place two outside "L" panels, move them to form a clean corner and skip shore the outside perimeter as necessary

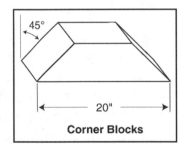

Corner Blocks

- Set top and bottom wales (or as many as necessary based on depth of trench). Wales should form a clean corner at the outside intersection of the corner panels. Anchor these in place by tying back to pickets

- Place "corner blocks" on wales using joist hangers or toe nails

- Shoot (feather shores into place then shoot) angled shores from inside "L" panel thrust blocks to corner blocks simultaneously

TIP: You should ensure that you feather and then shoot both shores from opposing sides at once to avoid kick.

Corner blocks used for L Trench

TIP: Each shore should have either one 23-degree or two 15- degree swivels. If using one 23-degree swivel shoot it to the corner block side. Never exceed more than 30 degrees on angles.

L Trench

- Repeat process for next set of shores

- Place and nail gusset plates on corner blocks

- Check and adjust shores

DEEP WALL TRENCH
15 FOOT TRENCH

DESCRIPTION OF TRENCH

Deep trenches are those trenches over 10 feet but not more than 15 feet. Deeper trenches present extreme forces that require commercial techniques when over 15 feet.

(Wales not shown to enhance the clarity of the diagram)

PROCEDURES

- Measure depth and width of trench

- Set pickets for tie backs (panels and wales)

- Prepare panels and wales

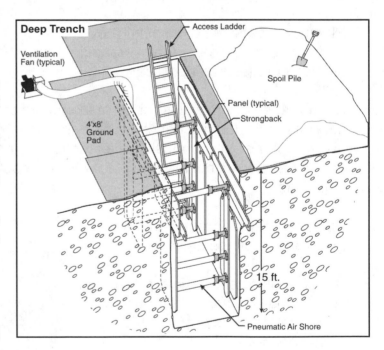

Deep Trench
Access Ladder
Ventilation Fan (typical)
Spoil Pile
4'x8' Ground Pad
Panel (typical)
Strongback
15 ft.
Pneumatic Air Shore

- Prepare shores

- Set deep wales on bottom floor of trench, picket back. If using aluminum wales and shore system, set in place at end of trench.

- Prepare all other wales, as necessary (one set every four feet). Use appropriate size timber for depth of trench.

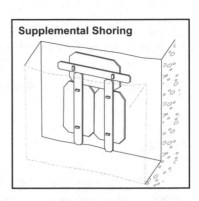
Supplemental Shoring

Deep trench panel placement

TIP: Each trench is different at this depth. You may find that you have to shoot panels to capture wall before placing wales. You must make this decision based on stability of the soil and your comfort level. If you do this you will have to maneuver the wales into place between shores.

- Lower bottom sets of panels in upright and tie back.

- Place top panels at 90 degrees to bottom panels, laying them lengthwise across two panels.

- Lower and place remaining sets of wales, not to exceed four feet vertical spacing between wales. Tie back to pickets.

- Shoot top panels using shore system. (No wales) Must be at least two feet from bottom of lip.

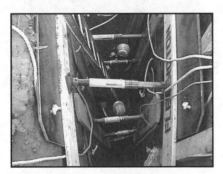
Horizontal top panels in a deep trench

- Shoot successive wale systems in place, working from top to bottom.

TIP: Each shore should have either one 23-degree swivel or two 15-degree swivels attached. When using one swivel shoot from the solid plate to the swivel side. NEVER EXCEED 30 DEGREES.

- At this depth "more is better." Add shores as necessary to control environment. A maximum of four foot vertical spacing shall be maintained.

- Check and adjust shores.

Summary

There are certain things that will need to be done on every trench regardless of the technique or system you build for stabilization. Eliminating all of the scene hazards is a very big job but crucial for the safety of your personnel. Establishing an Incident Management System that promotes organization, efficiency, and accountability is also vital to your overall success.

The area around the trench is very unstable and for that reason you will need to place ground pads. Once ground pads are in place you will want to monitor the trench for any atmospheric problems. Always remember to keep access to and from the trench a priority by establishing multiple escape routes for your personnel. For all below-grade work you will want ladder access every twenty-five feet.

It is very important to establish some method to document incident scene activities in case there is a cause to request reimbursement of expenses or the incident is litigated. Finally, you will want to critique all incidents to improve the efficiency of future trench rescues.

This unit concludes with a description of proper procedures for constructing a protective system in Straight Wall, Single Wall Slough, "T," and "L" trenches.

Questions (Answers and Discussion on page 262)

1. The minimum number of panels required for a T trench protective system is:

   a. 7
   b. 8
   c. 5
   d. 6

2. Your primary consideration during panel placement and shoring on a T-Trench is:

   a. Control the outer wall
   b. Assure wales are in place and tied back
   c. Capture and control the corners
   d. Set all panels at once

3. In the L-Trench, "thrust blocks" are required on all panels.

   a. True
   b. False

4. When constructing the L-Trench there are several force/counter force issues that need to be pre-planned for and engineered into the system. These include, but are not limited to:

   1. Panel kick at top and bottom
   2. Corner block kick
   3. Shore loosening
   4. Compressed water from the walls

   a. 1 and 4
   b. 1, 2, and 3
   c. 1 and 3
   d. 1, 3 and 4

5. What is the primary consideration when creating a system to render an L-Trench safe?

   a. Control the rear wall
   b. Control the inside corners

    c. Control the outside corners

    d. It doesn't matter; it's all a big cluster!

6. After the installation of pneumatic shores they should be toenailed to:

    a. Ensure a good set on the shore

    b. Keep the shore from falling out if it becomes loose

    c. Maintain the manufacturer's warranty on the shore

    d. All of the above are correct

7. Ventilation is required on all trench incidents.

    a. True

    b. False

8. In the single wall slough trench the technique to use would be:

    a. Inside wales

    b. Thrust blocks

    c. 7 panel set

    d. Outside wales

9. In order to create an open space in a trench the technique would be:

    a. Call for a trench box

    b. Use inside wales

    c. Use less shores

    d. None of the above

10. During deep trench operations (15-foot trench) it is acceptable to turn the top panels horizontal and shoot them first.

    a. True

    b. False

APPENDIX ONE

DEFINITIONS

TRENCH RESCUE

DEFINITIONS

ACCEPTED ENGINEERING PRACTICES: Those requirements that are compatible with standards of practice required by a registered professional engineer.

ACTIVE SOIL: The ability of the soil to contain energy as it relates to movement.

ALUMINUM HYDRAULIC SHORING: A pre engineered shoring system consisting of aluminum hydraulic shoring cylinders (cross braces), used with vertical rails (uprights) or horizontal rails (wales). Such a system is designed specifically to support the sidewalls of an excavation and prevent cave-ins.

ANGLE OF REPOSE: The natural angle at which loose particulate products will support their own weight, and which can be expected not to flow from a standing position.

AHJ: The acronym used to describe the authority having jurisdiction. Generally, this is the authority that certifies an individual competent in any particular rescue discipline.

BELL PIER: A type of shaft or footing excavation in which the bottom is larger than the cross section above to form a bell shape.

BENCHING: Modifying the walls of an excavation in such a way as to form horizontal steps with vertical faces, at predetermined angles and widths to prevent the soil from collapsing or sliding.

C- 60 SOIL: A class of soil that is a "moist, cohesive, or a moist dense granular soil" which does not fit into Type A or Type B classifications, and is not flowing or submerged. This material can be cut with near vertical sidewalls and will stand unsupported long enough to allow the shoring to be properly installed.

CAVE-IN: The separation of a mass of solid or rock material from the side of an excavation, or loss of the soil from under a trench shield or support system, and its sudden movement into the excavation, either by falling or sliding in sufficient quantity so that it could entrap, bury, or otherwise injure and immobilize a person.

CEMENTED SOIL: A soil in which a chemical agent similar to calcium carbonate holds together the particles, such that a hand size sample could not be crushed into powder or individual soil particles by finger pressure alone.

COHESIVE SOIL: Clay (fine grain soil), or soil with a high clay content, which has cohesive strength. Cohesive soil does not crumble, can be excavated with vertical side slopes, and is plastic when moist. Cohesive soil is hard to break up when dry, and exhibits significant cohesion when submerged. Cohesive soils include clay silt, sandy clay, silty clay, and organic clay.

COMPETENT PERSON: The individual, usually the supervisor or director of rescue operations who meets the OSHA standard for determining soil profiles, safety concerns, protective mechanisms and other requirements.

CONSENSUS STANDARDS: Standards developed by a group of persons who represent a particular industry, or product that is associated to that industry. The courts determine negligence but can use these standards that are not legally binding.

CROSS BRACES: The horizontal members of a shoring system installed perpendicular to the sides of the excavation, the ends of which apply pressure against either the uprights or wales.

DRY SOIL: Soil that does not exhibit visible signs of moisture content.

ENDS: The part of the trench where the walls meet the end.

EXCAVATION: An opening in the earth that is wider than it is deep.

FACES OR SIDES: The vertical or inclined earth surfaces formed because of excavation work.

FAILURE: The breakage, displacement, or permanent deformation of a structural member or connection that reduces its structural integrity and its supportive capabilities.

FISSURED: A soil material that has the tendency to break along definite planes with little resistance, or a material that exhibits open cracks, such as tension cracks in an exposed surface.

FLOOR: The bottom of the excavation.

GRANULAR SOIL: Gravel, sand, or silt with little clay content. Granular soil has no cohesive strength. Some moist granular soils exhibit apparent cohesion, but crumble when dry and cannot be molded.

GRAVITY: The gravitational attraction of the earth's mass for bodies at or near its surface.

HAZARDOUS ATMOSPHERE: An atmosphere that may be explosive, flammable, poisonous, corrosive, irritating, oxygen deficient, toxic, or otherwise harmful, and may cause injury, illness, or death.

HYDROSTATIC PRESSURE: Pertaining to trench rescue it is the pressure that results from the effects of water contained in soil.

KICK OUT: The accidental release or failure of a cross brace.

LEL: An acronym for Lower Explosive Limit, which represents the minimum concentration of product in air that will support combustion in the presence of a source of ignition.

LAYERED SYSTEM: Two or more distinctly different soil or rock types arranged in layers.

LIP: The area 360 degrees around the opening of the trench.

LOAM: A soil consisting of a friable mixture of varying proportions of clay, silt, and sand.

MOIST SOIL: A condition in which the soil looks and feels damp. Moist cohesive soil can be easily shaped into a ball and rolled into small diameter threads before crumbling. Moist granular soil that contains some cohesive material will exhibit signs of cohesion between particles.

OSHA: The Occupational Safety and Health Administration that is a Federal office and also an agency in some states.

PCF: The acronym that describes the term pounds per cubic foot.

PPE: The acronym that describes the term personal protective equipment.

PASSIVE SOIL: That characteristic that describes a soil with no potential for movement.

PLASTICITY: The property that allows the soil to be deformed or molded without appreciable change in total volume.

PROTECTIVE SYSTEMS: Pre-engineered systems or system components designed to protect construction employees or rescue personnel from cave-ins, collapses, falling material, and other equipment.

RAMP: An inclined walking or working surface used to enter one point from another, and constructed from earth or other structural materials such as steel or wood.

REGISTERED PROFESSIONAL ENGINEER: A person who is officially registered as a "professional engineer" in the state where they are working. However, a "professional engineer" registered in any state is deemed to be a "registered professional engineer" within the meaning of this standard when approving designs for manufactured protective systems or tabulated data to be used in interstate commerce.

SABA: The acronym used to describe the supplied air breathing apparatus. The SABA system is used to provide a remote supply of air to the rescuer or victim for respiratory purposes.

SCBA: The acronym used to describe the self-contained breathing apparatus. It is used to supply air from a system worn by the rescuer and does not require a remote source of air.

SATURATED SOIL: A soil in which the voids are filled with water. Saturation does not require flow. Saturation or near saturation is necessary for the proper use of pocket penetrometer or shearvane soil testing devices.

SHEETING: These could be interconnected uprights, sheets of timber or shore-form panels used in contact with the walls of the trench. They function as a shield system with uprights, wales, and other engineering systems.

SHIELD (SHIELD SYSTEM): Permanent or portable structures, such as trench boxes, rabbit boxes, coffins, etc., designed to withstand the forces of a collapse or cave-in.

SHORE: Horizontal members, installed perpendicular to the wall of a trench, whose ends press against the uprights, wales, or panels to create pressure zones and support.

SHORING: A system made of timber, metal, hydraulic, or mechanical members that support the walls and prevent cave-ins. Used to support sheeting in conventional rescue operations.

SILT: An earth matter comprised mostly of sand that is carried by water and deposited as sediment.

SLOPING (SLOPING SYSTEM): Excavating the walls so that they incline away from the excavation at a predetermined angle according to the soil profile to prevent cave-in or soil movement.

SPOIL PILE: The dirt taken out of the trench and piled along the side of the trench. The spoil pile must have at least a two-foot setback from the trench lip.

STABLE ROCK: A natural solid material that can be excavated with vertical sides and will remain intact when exposed. Unstable rock is considered stable when the rock material on the sides of the excavation is secured against movement by rock bolts, or another protective system designed by a registered professional engineer.

STRUCTURAL RAMP: A ramp built of steel or wood, used for vehicular access. Ramps made of soil or rock are not considered structural ramps.

SUPPORT SYSTEM: A structure such as underpinning, bracing, or shoring, which provides support to an adjacent structure, underground installation, or sides of an excavation.

TABULATED DATA: Tables and charts representing information approved by a registered engineer and used to design and construct a protective system. These are found in the shoring tables in the OSHA manual, and may be constructed by your own engineer but must be predicated on "pre engineered data."

TSF: The acronym for the term tons per square foot.

TOE: The area where the walls and floor intersect at a 90-degree angle at the bottom of the trench and two feet up the walls.

TRENCH: An opening in the ground that is deeper than it is wide.

TRENCH BOX: See "shield."

TYPE A SOIL: A soil with an unconfined compressive strength of 1.5 tons per square foot or greater.

TYPE B SOIL: A soil with an unconfined compressive strength of greater than 0.5 tons per square foot, but less than 1.5 tons per square foot.

TYPE C SOIL: A soil with an unconfined compressive strength of less than 0.5 tons per square foot.

UNCONFINED COMPRESSIVE STRENGTH: The force or load per unit area, as calculated with a penetrometer or other device, and stated numerically in tons per square foot that determines the point at which a soil will fail in compression.

UPRIGHTS: Vertical members placed in contact with the walls, or panels (sheeting), that may or may not contact each other. More than one upright may be used on each sheeting system.

WALES: Horizontal members of a shoring system placed parallel to the walls and whose sides bear against the uprights or the excavation shoring system or face. Wales can be 6" x 6", 8" x 8", or 10" x 10" wood timbers or comprised of various steel and aluminum components.

WALLS: See "Faces" or "Sides."

WET SOIL: A soil that contains more moisture than moist soil, but is in such a range of values that the cohesive material will slump or begin to flow when vibrated.

APPENDIX TWO

PRACTICAL SITE AND
EQUIPMENT REQUIREMENTS

TERMINAL OBJECTIVE

To provide the student and the instructor with the
necessary information to construct trench rescue mock-ups
and to specify the necessary equipment needed to teach
operations and technician level trench rescue classes.

PRACTICAL SITE REQUIREMENTS

AWARENESS: Suitable area for trench rescue equipment to be set up for demonstration and familiarization purposes.

OPERATIONS: Two types of trenches

TIP: When training, always slope one end of trench to allow an escape path that does not require ladder egress.

Single Wall

20 ft

5 ft

Single Wall Slough

4 ft

20 ft

5 ft

5 ft

8 ft

5 ft

20 ft

4 ft

16 ft

5 ft

20 ft

TECHNICIAN: Intersecting "T" and the "L" trench with place for Deep Wall.

TRENCH RESCUE RESOURCE LIST

Two Trenches for Maximum of 30 Students

- All trenches dug to specifications as indicated on previous diagram. Specifications provided for straight wall trench, single wall slough, L-trench, and T-trench.

- Each class has a different number of trenches depending on the number of students and the type of class.

- Backhoe on-site for class (if possible, an excavator would also be beneficial).

- Fourteen 4' x 8' panels with 2" x 12" x 12' strong backs. These can be home made or can be Shorform Panels (preferred).

- Six 2" x 12" x 12' lumber

- Ten 2" x 4" x 12' lumber

- Four 4" x 4" x 12' timbers

- Ten 4" x 6" x 10' timbers

- Four 6" x 6" x 14' timbers

- Sixteen 4' x 8' x 1/2" sheets of exterior plywood (ground pads / supplemental sheeting)

- Twenty-five pounds 16-penny duplex nails

- Twelve 20 ft. sections of utility rope (7/16" or 1/2" diameter)

- Ten steel pickets

- Six carpenter nail pouches

- Six claw framing hammers

- Six 25' tape measures

- Two 4-pound hand mauls

- Two 8-pound sledge hammers

- One chain saw and support equipment (tools, gas, oil, extra chain, etc.)

- Two portable generators of sufficient wattage to run your equipment and power cords

- Two ventilation fans

- One diaphragm pump (Mud Hog) de-watering device, with intake and discharge hoses

- One centrifugal pump dewatering device, with intake and discharge hoses

- Four long-handle shovels (flat and round point)

- Two steel yard rakes

- Four small entrenching tools

- Six 5-gallon buckets

- Two rescue mannequins or hose dummies

- Two patient packaging devices: Stokes, LSP Halfback, Miller Body Splint, etc.

- One three-function atmospheric monitor

- Sixty pieces 4" x 4" x 15" cribbing

- Twenty wedges, cut from 4" x 4" x 15"

- Two 20-foot Nylon rigging straps

- Six large locking carabiners

- One complete set of low-pressure air bags

- One complete set of high-pressure air bags

- One of the following sets of additional shores (preferably all of them)

  - Two sets of hydraulic shores
  - Twelve B/C Air Shores with swivels, extensions, air cart and hoses
  - Six screw jacks (1 1/2" or 2" pipe)
  - One Paratech Struts Trench Kit

- One section 10-foot steel pipe or concrete pipe 12"-18" diameter

- Rehab supplies: water, Gatorade, cups, etc.

- Fire line tape

- Access to air cascade system to fill bottles

- Extra SCBA cylinders

- Clean up area and supplies (brushes, hose, soap, towels, oil, rubber lube, etc.—everything needed to get your equipment back in shape after final usage)

TIP: You would be surprised at the number of internal and external resources that you have to obtain these items. Examples include:

- Departmental fire and rescue
- Public Works, Public Utilities (city, county, or private)
- Local vendors

APPENDIX THREE

SAMPLE TRENCH RESCUE STANDARD
OPERATING PROCEDURE

TERMINAL OBJECTIVE

To provide the student with a sample trench rescue
standard operating procedure.

TRENCH AND EXCAVATION
COLLAPSE PROCEDURES

PURPOSE

To address operations which involve the location, disentanglement, and removal of victims from underground collapses in trenches and excavations.

SCOPE

This policy is designed to provide guidelines to all fire and EMS units when presented with an incident involving the collapse of a trench or excavation where a victim(s) is (are) trapped or buried. This includes "protected" trenches where a victim(s) is (are) trapped or pinned by heavy equipment, pipe, bedding material or items other than soil.

DEFINITION

As defined by the OSHA regulation 29CFR Part 1926:

TRENCH: A narrow excavation in relation to its length made below the surface of the ground. In general, the depth is greater than the width, but the width is not greater than 15 feet.

EXCAVATION: A man-made cut, cavity, trench or depression in an earth surface, formed by earth removal. Usually wider than it is deep.

1. GENERAL GUIDELINES

A. Any incident that involves a trench or excavation shall be the responsibility of the Special Operations/Technical Rescue Team.

B. Any incident in which a victim is trapped, buried or experiencing a medical emergency in a trench or excavation will require the response of the Technical Rescue Team.

C. No firefighter or EMS person shall enter an unprotected trench to render victim care or perform disentanglement operations. All trenches shall be "safe and protected" using approved methods prior to entry by any emergency personnel.

D. All emergency vehicles shall park at least 100 feet from the collapse site. The only exception to this shall be the technical rescue team vehicle, which may park no closer than 50 feet.

E. All traffic within 300 feet of the collapse zone shall be stopped or detoured.

F. A hazard zone shall be established to control at least 75 feet around the perimeter of the collapse zone. This should be done with fire line tape.

OPERATIONAL PHASES: FIRST DUE UNITS

I. ASSESSMENT

A. First due units should attempt to gather the following information:

1. What is the nature of the problem? Collapse, trap, medical, etc.
2. How many victims are there?
3. What is their location?
4. Width, length, and depth of the trench?
5. Are there any on-scene hazards:
   a. Disrupted utilities
   b. Flowing water
   c. Secondary collapse
   d. Mechanical hazards/heavy equipment
   e. Exposed but non-disrupted utilities
   f. Hazardous materials/explosives

B. Once these items are evaluated, the following should be completed:

1. Determine whether the incident involves a rescue or a recovery.
2. Ensure Technical Rescue Team response and full assignment.
3. Establish visible command and control access to collapse area.

II.   MAKING THE SITE SAFE

A. General area safety

   1. Protect the general area around the collapse zone for at least 300 feet in all directions. This includes:

     a. Traffic control

     b. Access control

     c. General hazard identification

     d. Shutting down all heavy equipment.

B. Rescue Area Safety

   1. "Perform" or "conduct" the initial steps needed to make the actual collapse zone around and in the trench as safe as possible using basic techniques. SHEETING AND SHORING OPERATIONS, ENTRY AND DISENTANGLEMENT OPERATIONS SHOULD BE CARRIED OUT UNDER THE DIRECTION OF THE TECHNICAL RESCUE TEAM.

     a. Ground pad the trench or collapse site lip.

     b. Ventilate the trench with Positive Pressure Ventilation, if needed.

     c. Support any unbroken utilities

     d. If medical conditions permit, provide a helmet and goggles for the victim (preferably not a fire service helmet).

     e. DO NOT ALLOW ANY PERSONNEL INTO AN UNPROTECTED TRENCH.

     f. DO NOT TOUCH OR LEAN ON ANY HEAVY EQUIPMENT UNTIL YOU HAVE ENSURED IT IS NOT IN CONTACT WITH ELECTRICAL UTILITIES!

STOP: **AWAIT THE ARRIVAL OF TECHNICAL RESCUE TEAM PERSONNEL AND EQUIPMENT.**

TECHNICAL RESCUE OPERATIONS PHASE

I. OPERATIONAL RESPONSIBILITY

A. All personnel shall report to and work through the incident command post.

B. Establishment of sector officers associated with the trench or excavation collapse may be necessary. These shall be in accordance with the Incident Command Policy.

C. In some cases the following Sector Officers shall be established:

1. OPERATIONS: Responsible for coordination of the actual collapse site and the sectors associated with all activity in the "rescue area."

2. EXTRICATION: Responsible for directing the actual sheeting and shoring, disentanglement and removal operations associated with the trench or excavation. Operations personnel will report directly to the Operations Officer.

II. COLLAPSE ZONE OPERATIONS

A. Different collapse scenarios will obviously require different sheeting and shoring techniques as the situation demands. Each scenario should be evaluated using the same evaluation mechanism and adaptations made to the current operation as required by the configuration of the trench or excavation.

B. The following are potential forms of collapse which may be encountered. They should be handled in accordance with accepted techniques previously taught.

1. Single Wall Sheer
2. Double Wall Sheer
3. Spoil Pile Slide
4. Intersecting Trench Collapse
5. Collapses in "Protected Trenches"
   a. Rabbit Box slide or above level collapse
   b. Industrial shoring collapse
   c. Inadequate protection systems in place

C. The following are potential forms of victim entrapment scenarios that may be encountered:

1. Victim(s) buried to waist
2. Victim(s) buried to chest
3. Victim(s) not buried but injured or experiencing a medical problem in the trench environment
4. Victim(s) trapped or pinned by heavy equipment or pipe
5. Victim(s) trapped in running sand or material
6. Victim(s) completely buried
7. Victim(s) buried in the end of a large diameter pipe

III. OPERATIONAL GUIDELINES

A. RESCUE AREA CONSIDERATIONS:

1. Ensure ventilation continues when needed and monitor atmosphere
2. Ensure de-watering systems are operational
3. Ensure utilities are controlled and identified
4. Limit personnel at lip and collapse zone
5. Ensure communications with logistics area via department radio system or landline
6. Ensure Safety Officer is in control of access and personnel
7. Ensure media staging area is located away from the collapse zone

B. GENERAL CONSIDERATIONS:

1. Brief all personnel on plan of action and confer with appropriate sectors
2. Provide constant updates to Incident Command
3. Plan at least two steps ahead of the operation; have a secondary plan ready in the event that the initial tactical plan proves unworkable
4. Rotate personnel regularly
5. Ensure personnel involved in disentanglement and digging operations are rotated at least every 30 minutes

C. VICTIM CONSIDERATIONS:

1. ABOVE ALL, TREAT VICTIM FOR CRUSH SYNDROME IN ACCORDANCE WITH PROTOCOLS
2. Consider and treat for hypothermia

3. Never dig a victim out with heavy equipment
4. Once you are proximate to the victim, dig by hand
5. Consider the use of helicopter transport to a trauma center
6. Ensure Technical Rescue Team Paramedics coordinate and direct victim packaging operations at all times
7. Plan movement mechanism well ahead of time for the removal of the victim once disentangled

D. COMMUNITY RESOURCES:

1. In the event that Public Utility resources are needed, advise the following:

    a. Exactly what is needed:

        1.   Manpower
        2.   Heavy equipment (what kind)
        3.   Pumps (what type)
        4.   Vac truck

    b. Ensure that all identified utilities have a representative present. DO NOT ATTEMPT TO CONTROL UTILITIES.

    c. Ensure a staging area for all incoming community resources requested

SPECIAL SITUATIONS:

A. RUNNING SAND OR MATERIAL

1. In these cases, it may be necessary to encase the victim(s) in interlocking drums used as an isolation tunnel. Remember that in all cases these drums should be used IN THE VERTICAL POSITION ONLY. USING THEM HORIZONTALLY MAY CAUSE THEM TO FAIL AND CRUSH UNDER THE WEIGHT OF MATERIAL.

B. Other items which may be used for isolation tunnels in either the vertical or horizontal configurations are:

1. Concrete or steel pipe
2. Corrugated pipe

C. PIER HOLES OR CAISSONS

1. These are bell shaped excavations, which are used mainly as "footers" to pour support columns for concrete buildings. They represent extreme danger due to the difficulty in sheeting and shoring and their bell shaped bottoms. Extreme caution should be exercised when involved in these types of operations.

D. TRENCH AND TUNNEL OPERATIONS

1. In certain cases it may be necessary to dig a parallel trench or excavation in order to create a parallel shaft. If this becomes necessary, consider the following:

   a. Any trench cut for a rescue operation should be properly protected by either conventional or industrial means.
   b. Ensure all utilities are identified prior to cutting the trench. This can be done by requesting the utility company(ies) on an emergency basis.
   c. Assure adequate shaft material for construction of your parallel shaft.
   d. If possible, request and retain a certified engineer to assist in the planning and implementation.
   e. This should be used only as a last option.

TERMINATION

A. Rehab all personnel prior to termination and removal operations.

B. Brief all personnel on the operation and its intended outcome.

C. Perform removal operations in the REVERSE ORDER.

D. Beware of secondary collapse zones. No equipment is worth risking an injury.

E. Stage, clean, and inventory all equipment. Report any lost or damaged equipment.

F. Any parallel shaft construction, tunnels, or isolation tunnels should be left in place. Removing them may cause a collapse.

APPENDIX FOUR

TRENCH RESCUE TACTICAL WORKSHEETS

TERMINAL OBJECTIVE

To provide the student with a sample trench rescue tactical worksheet.

TRENCH RESCUE TACTICAL WORKSHEET

DATE	SHIFT
ADDRESS	TIME
OWNER	WIND SPEED
RESPONSIBLE PARTY	DIRECTION
RESPONDING COMPANIES	
BACKFILL COMPANIES	

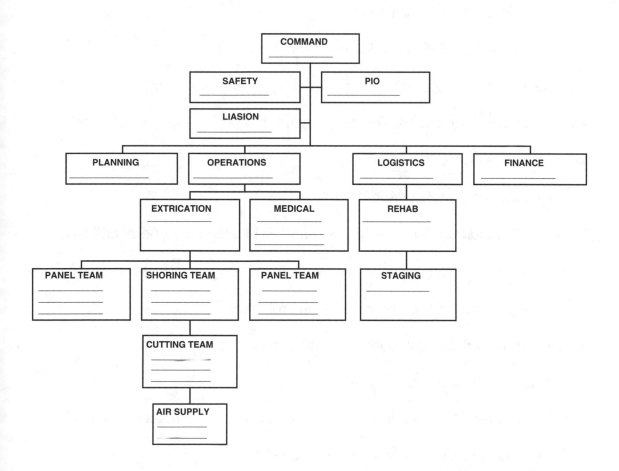

TRENCH RESCUE SITUATION SIZE-UP QUESTIONS

What type of collapse has taken place?

_____

_____

What has occurred prior to my arrival?

_____

_____

Do I have sufficient manpower?

Is the IMS flow chart expanded sufficiently to handle the incident?

Are all necessary IMS positions filled?

Are rescue and/or EMS transport units on scene or en route?

Do I have sufficient specialized equipment on scene or en route?

Do I need to call for specialized civilian personnel (e.g. engineer or rigger)?

Is this a rescue or recovery?

Is the victim completely buried?

What is the long term weather forecast and will it affect my operation?

Have I eliminated all hazards in the area? General_____   Rescue area_____

Are there any hazardous materials involved?

Have I monitored the atmosphere in and around the trench?

Is ventilation needed and in place?

Have I determined the correct protective system to make rescue attempt?

Do I need to consider commercial techniques?

Have I assigned a Public Information Officer to handle the media?

SKETCH OF TRENCH RESCUE SCENE

SIDE C

SIDE B

SIDE D

SIDE A

NOTES:

APPENDIX FIVE

NFPA 1670, Chapter 9

TRENCH AND EXCAVATION

Reprinted with permission from NFPA 1670, *Operations and Training for Technical Rescue Incidents* Copyright © 1999, National Fire Protection Association, Quincy, MA 02269. This reprinted material is not the complete and official position of the National Fire Protection Association on the referenced subject, which is represented only by the standard in its entirety.

Chapter 9
Trench and Excavation

9-1 General Requirements.

9-1.1 Organizations operating at trench and excavation incidents shall meet all the requirements specified in Chapter 2 of this standard.

9-1.2* The AHJ shall evaluate the effects of severe weather, extremely hazardous trench or excavation situations, and other difficult conditions to determine whether their present training program has prepared the organization to operate safely.

9-2 Awareness

9-2.1 Organizations operating at the awareness level shall meet all requirements specified in Section 5-2 within this standard, the requirement in Chapter 2 of NFPA 472, *Standard for Professional Competence of Responders to Hazardous Materials Incidents,* and the requirements of *competent person* as defined in Section 1-3 of this standard.

9-2.2 Awareness-level functions at trench and excavation emergencies shall include the following:

(a)*   Size-up of existing and potential conditions
(b)*   Identification of the resources necessary to conduct safe and effective trench and excavation emergency operations
(c)*   Development and implementation of procedures for carrying out the emergency response system for trench and excavation emergency incidents
(d)*   Development and implementation of procedures for carrying out site control and scene management
(e)*   Recognition of general hazards associated with trench and exca-

vation emergency incidents and the procedures necessary to mitigate these hazards within the general rescue area

(f)* Recognition of typical trench and excavation collapse patterns,
the reasons trenches and excavations collapse, and the potential
for secondary collapse

(g)* Development and implementation of procedures for making a rapid,
non-entry extrication of non-injured or minimally injured victim(s)

(h)* Identification of the resources necessary to conduct safe and effective trench and excavation emergency operations

9-3 Operations.

9-3.1* Organizations operating at the operations level shall meet all requirements specified in Section 9-2. In addition, members shall be capable of
hazard recognition, equipment use, and techniques necessary to operate safely and effectively at trench and excavation emergencies, including the collapse or failure of individual, nonintersecting trenches with an
initial dept of 8 ft (2.44m) or less where no severe environmental conditions exist, digging operations do not involve supplemental sheeting and
shoring, and only traditional sheeting and shoring are used.

9-3.2 Organizations operating at the operations level shall meet all requirements specified in Sections 4-3, 5-3, and 6-3.

9-3.3 Operations-level functions at trench and excavation emergencies shall
include the following:

(a) Development and implementation of procedures to make an entry
into a trench or excavation rescue area

(b)* Recognition of unstable areas associated with trench and excavation emergencies and adjacent structures

(c)* Development and implementation of procedures to identify probable victim locations and survivability

(d)* Development and implementation of procedures for making the
rescue area safe, including the identification, construction, application, limitations, and removal of traditional sheeting and shoring
using tabulated data and approved engineering practices

(e)* Development and implementation of procedures for initiating a one-
call utility location service

(f)* Identification of soil types using accepted visual or manual tests

(g)   Development and implementation of procedures to ventilate the trench or excavation space

(h)   Identification and recognition of a bell-bottom excavation (pier hole) and its associated unique hazards

(i)   Development and implementation of procedures for placing ground pads and protecting the "lip" of a trench or excavation

(j)*   Development and implementation procedures to provide entry and egress paths for entry personnel

(k)*   Development and implementation procedures for conducting a pre-entry briefing

(l)*   Development and implementation of procedures for record keeping and documentation during entry operations

(m)*   Development and implementation of procedures for implementing and utilizing a rapid intervention team (RIT) as specified in Section 6-5 of NFPA 1500, *Standard on Fire Department Occupational Safety and Health Program*

(n)   Development and implementation of procedures for the selection, utilization, and application of shield systems

(o)*   Development and implementation of procedures for the selection, utilization, and application of sloping and benching systems

(p)   Identification of the duties of panel teams, entry teams, and shoring teams

(q)   Development and implementation of procedures for assessing the mechanism of entrapment and the method of victim removal

(r)*   Development and implementation of procedures for performing extrication

9-4   Technician.

9-4.1* Organizations operating at the technician level shall meet all requirements specified in Section 9-2 and 903. In addition, members shall be capable of hazard recognition, equipment use, and techniques necessary to operate safely and effectively at trench and excavation emergencies, including the collapse or failure of individual or intersecting trenches with an initial depth of more than 8 ft (2.4m) or where severe environmental conditions exist, digging operations involve supplemental sheeting and shoring, or manufactured trench boxes and/or isolation devices would be used.

9-4.2 Organizations operating at the technician level shall meet all requirements specified in Section 5-4 and 6-4.

9-4.3 Technician-level functions at trench and excavation emergencies shall include the development and implementation of the following:

(a)* Procedures for the identification, construction, application, limitations, and removal of manufactured protective systems using tabulated data and approved engineering practices

(b)* Procedures to continuously, or at frequent intervals, monitor the atmosphere in all parts of the trench to be entered. This monitoring shall be done in the following order, for oxygen content, flammability (LEL/LFL), and toxicity

(c) Procedures for the identification, construction, application, limitations, and removal of supplemental sheeting and shoring systems designed to create approved protective systems

(d) Procedures for the adjustment of protective systems based on digging operations and environmental conditions

(e)* Procedures for rigging and placement of isolation systems

APPENDIX SIX

ANSWERS AND DISCUSSIONS

FOR UNIT QUESTIONS

Unit One   Considerations for Specialized Operations

Answers

1. The term big three refers to specialized rescue operations that require:

    d.   Special people, special equipment, and special training

Discussion: In order for a team to become proficient in trench rescue a commitment to recruit special people, purchase special equipment, and provide special training is necessary. A failure in any leg of the "big three" will decrease the team's chances of a successful and safe trench rescue operation.

2. The service level(s) as identified by NFPA 1670 that apply to rescue operations and may be an indicator of level of competency are Awareness, Operations, Technician, and Instructor.

    b.   False

Discussion: The 1999 Edition NFPA 1670 identifies the three service levels of Awareness, Operations, and Technician. For the purpose of technical rescue training the Instructor level is not only certified at the Technician level but also maintains the skills to teach and instruct.

3. Operations level personnel are trained to:

    c.   The first level at which personnel learn the necessary techniques to work in trenches

Discussion: Operations level trench personnel are the first level of personnel who can actually enter non-intersecting trenches of 8 feet or less in depth. Awareness personnel are trained to identify hazards and begin incident stabilization as well as non-entry rescue of persons without entering the trench. Technician level personnel, in addition to operating in non-intersecting trenches, can provide rescue efforts in intersecting trenches and deep trenches.

Unit Two   **Trench Rescue Decision Making**

Answers

1. The primary factor used in performing a risk/benefit analysis is:

     c. Is the event a rescue or recovery

Discussion: If the trench incident is a true rescue, then you are involved in a true emergency. In these situations we assume a higher level of risk based on the potential or probable benefit of saving the victim. In no case should you assume risk to recover a dead body.

2. The "L" in the failure acronym describes:

     d. Lack of teamwork and experience

Discussion: The true success of a team is told in their ability to work together for the common good. Selfless team members understand that there are many jobs that need to be done at a trench rescue and they are not always going to get the best one. True team players do not let this deter their motivation to do a good job.

3. Most completely buried victims ultimately survive a trench collapse.

     b. False

Discussion: Unless a victim of a collapse was able to shield himself in a pipe or other protective box, it is very unlikely that the completely buried victim will survive a trench collapse. Your risk/benefit analysis should include the probability of survival when making decisions regarding a rescue plan.

Unit Three   **Preparing the Rescue System**

Answers

1. There are a number of different methods to move and store your equipment. Which of the following would be least appropriate:

     d. Vehicle rescue truck

Discussion: The type of transportation system you choose for the team's trench rescue equipment should be based on the amount, size, and weight of your equipment. Additionally, the vehicle should be appropriate for the money you have to support its maintenance.

2. The most reliable form of rescue team development is the:

    c. Self-sufficient

Discussion: Being a self-sufficient team means you fully control the ability to provide personnel, training, and equipment for trench rescue. It is also the method that costs the most to maintain and requires the greatest commitment on the part of the sponsoring organization.

3. By far the most important aspect concerning the potential success of your team will be:

    b. The people you choose for your team

Discussion: Talented, organized, and committed team members are the backbone of the rescue system. In most cases, good team members can overcome many potential problems with the rescue system such as difficult and unusual rescue situations or lack of proper equipment.

Unit Four   Introduction to Trench Rescue

Answers

1. Statistics show that trench accidents have a ___ fatality rate than other types of construction accidents:

    a. Higher

Discussion: As you will study in Unit Six, soil and the physical forces associated with collapses are very unforgiving. Most completely or significantly buried victims end up as trench fatalities.

2. The OSHA standard for trenches and excavations is:

    a. OSHA CFR 1926 Subpart P

Discussion: Occupational Safety and Health Administration Code of Federal Regulations 1926 Subpart P is the construction standard for excavations.

3.   On the scene of a trench or excavation you could expect to find a person who is familiar with all aspects of soil types and testing called:

d.   Competent person

Discussion: As required by the OSHA standard, this person would be on scene to identify hazards and approve the protective system place based on the soil profile.

4.   A trench is an excavation that is generally deeper than it is wide but its width measured at the bottom does not exceed 15 feet.

a.   True

Discussion: By definition, every depression cut in the earth is an excavation; however, for the purpose of the OSHA standard and trench rescue operations, this is the most applicable definition.

5.   Egress ladders in a trench must be within ___ feet of a worker in a protected trench:

d.   25

Discussion: While the standard requires a method of escape every 25 feet, it is advantageous to always provide a means of escape from both sides of any trench. The method of escape could include a sloped ramp on one or both ends.

6.   The overriding reasons for contractor non-compliance with trench protective systems is:

a.   Time and money considerations

Discussion: Trench protective systems take time to set up and put in place. Generally, the faster a contractor can work the more money they will make.

7.   The minimum setback requirements for the excavated spoil pile:

b.   2 feet

Discussion: For rescue purposes, the farther away from the trench lip the exca-vated soil can be placed the least amount of compressive forces it will add to the trench walls.

8. The OSHA standard on trenches and excavations is :

    e. Both b and d

Discussion: The OSHA CFR 1926 Subpart P is a construction standard that is performance based. Performance-based standards allow for deviation from the specific requirements of the standard if the proposed protective system meets or exceeds the original protective system requirements.

9. The OSHA standard was originally part of the:

    a. Contract Work Hours Standard Act

Discussion: The original Contract Work Hours Standard Act was the first of the construction standards for excavations. It was difficult to understand and the protective system requirements were generally more expensive to build than the fines for not using them.

10. The height of the spoil pile is taken into account when determining the need for a protective system:

    a. True

Discussion: The height of the spoil pile is always taken into consideration when determining the need for a protective system.

Unit Five   Trench Incident Management & Support Operations

Answers

1. The Operations Officer at a trench collapse:

    e. All of the above

Discussion: The Operations Officer at a trench collapse needs to understand how the many different components of the rescue operation fit together. Several types of activities are ongoing and still others are very intensive but for only short periods of time. Managing the factors involved in assigning limited resources is critical to the success of the operation, and therefore this individual needs to be one of your most experienced officers.

2. The minimum number of persons that should be assigned to a panel team is:

    b. 4

Discussion: Panel work is very labor intensive work. Panels are heavy and awkward to maneuver in and around an open trench and for that reason four persons should be assigned to a panel team. In manpower critical situations fewer people can be used; however, the reduced number of members are susceptible to injury and fatigue.

3. Which of the following is not a critical IMS position assignment when initially getting started at the scene of an incident?

    b. Rehab Officer

Discussion: This is not to say that rehabilitation is not important. However, you will find that in short to medium time incidents, rescue folks run on adrenaline and will have to be forced to take a break. Consider establishing a rehab sector and make rescuers take periodic breaks to hydrate.

4. A Rapid Intervention Team (RIT) should always be established on the scene of a trench collapse operation:

    a. True

Discussion: NFPA 1500 Standard on Fire Department Health and Safety describes the use and necessity of Rapid Intervention Teams. It is also a requirement of the 1999 NFPA 1670 Chapter 9 Trench and Excavation standard.

Unit Six   **Soil Physics**

Answers

1. For the purpose of determining total soil weight, we can estimate that a cubic foot of soil weighs approximately:

    b.   100 lbs. pcf

Discussion: Soil weight is a sum of the water and rock contained in a given sample. The more water added to the soil the heavier the soil will be in weight.

2. The most dangerous portion of the unprotected trench wall is:

    a.   Just below the middle

Discussion: The bottom of a trench has the greatest amount of pressure; however, the angle of the trench at the toe adds a measure of support. Generally, failure occurs at a point on the trench wall where the angle at the bottom has lost its ability to provide support when compared to the pressure from the top.

3. The term _____ best describes the effects of mass as it relates to the tendency of one object to be attracted to another object.

    b.   Gravity

Discussion: Gravity in most cases is the enemy of the rescuer. By shoring a trench its forces are managed by transferring the energy from one side of a trench to the other.

4. The amount of resistance soil has to pressure is a measurement of:

    a.   Unconfined Compressive Strength

Discussion: As we will discuss in Unit Nine on Soil Classification and Testing, the UCS of the trench wall is a measurement of resistance as recorded by a soil testing instrument like a penetrometer, shearvane, or torvane.

5. The effects of water can add a tremendous amount of weight to the exca-
    vated material because water weighs approximately:

    d. 62.4 pcf

Discussion: Per cubic foot is a measurement of volume and should not be confused with pounds per square foot or inch, which is a measurement of pressure.

Unit Seven   Conditions & Factors that Lead to Collapse

Answers

1. Which of the following can be considered a factor that can contribute to a
    trench collapse:

    e. All of the above

Discussion: It is good practice to eliminate all of the conditions that can possibly
lead to a collapse prior to initiating any trench rescue activities.

2. A good indication that soil has been previously disturbed is:

    e. B and C are correct

Discussion: If you see installed utilities in the trench, foreign matter in the trench
walls, or spoil pile, the soil has most likely been previously disturbed.

3. You should always move heavy equipment from near the trench:

    b. False

Discussion: Moving equipment that is near the trench may cause additional
weakening of the trench environment; however, in some cases the equipment
will prevent an adequate shoring system from being assembled. It is always
best to leave well enough alone if that is an option, but in all cases, someone
should disable the equipment by removing the keys and keeping them in a
secure location.

4. Vibration is a factor that can cause a trench collapse and can be attributed to:

    d. All of the above

Discussion: There are many types of emergency and non-emergency events that can cause vibration. All of these potential factors need to be evaluated and addressed during your size-up of the incident.

5. Of the factors that can contribute to a trench collapse, the amount of time the trench is open is a major factor. The time the trench is open is:

    d. Freestanding time

Discussion: Freestanding time allows many of nature's elements like rain and wind to weaken the trench walls.

Unit Eight   Types of Trench Collapses

Answers

1. When the excavated material falls back into the trench it is best described by the term:

    b. Spoil pile slide

Discussion: Occurs when excavated soil falls back into the trench.

2. A collapse in which the material loses its ability to stand and fails along a mostly vertical plane best describes a:

    a. Shear wall collapse

Discussion: Occurs frequently after a trench has had freestanding time and been exposed to the elements. This type of collapse can be very large sections of earth.

3. The most common type of soil failure in an intersecting trench is:

    a. Wedge

Discussion: This is the result of two sides of an intersecting trench wall not being supported.

4. A condition that presents itself as a failure at the toe and on both sides of a trench is:

    d. Bell pier condition

Discussion: A bell pier condition usually results from a sandy or water saturated soil in the bottom of the trench.

Unit Nine   Soil Classification and Testing

Answers

1. When classifying soils, which of the following is not an approved testing method?

    d. Water drip test

Discussion: There is no such thing as a water drip test. Other methods described are approved as either visual or manual testing methods.

2. A test that is performed by rolling the soil threads, which determines the soil's ability to be deformed and molded, best describes:

    d. Plasticity test

Discussion: This test determines a soil's cohesiveness and propensity to stay together. This is one of the tests that best indicates the soil's ability to stand and therefore its future collapse potential.

3. Warning signs of potential collapse consist of:

    d. All of the above

Discussion: Cracks in the trench lip or exposed walls, as well as loose and falling soil, are all indicators of potential collapse. In addition, the amount of freestanding time a trench has been exposed could indicate excessive drying of the soil and therefore loss of strength.

4. You are performing a quick evaluation of the soil at a collapse operation and notice that the excavated material is dry, cracked, and granular. As a manual test you attempt to roll the soil in your hand which causes it to break and crumble. Based on your assessment this would be best classified as:

    c. Type C

Discussion: Soil that is easy to crumble and dry will most likely contain sand. This type of soil can look very stable when it is moist but lose almost all of its strength when dry. The manner in which the excavated material crumbles and its dryness would most likely indicate Type C soil.

5. The trench you are examining is 10 feet deep and four feet wide with soil that is a clay loam mixed with sand. The upper portions of the trench have water seeping from the walls with an additional two inches of water in the bottom of the trench. Based on your assessment you conclude that this type of soil is:

    c. Type C

Discussion: Any condition of a trench that presents as running or seeping water will almost always be considered Type C soil.

6. Cohesive materials that have an unconfined compressive strength greater than .5 tsf but less than 1.5 tsf best describes:

    d. Type B

Discussion: The answer is the unconfined compressive strength described as the numerical range determined by OSHA to represent Type B soil. Other factors determined in the examination may indicate Type C soil even if the soil has the range indicated for Type B.

7. The test that is used to determine a soil's propensity to fissure is :

    e. None of the above

Discussion: The dry strength test is used to determine a soil's propensity to fissure. Breaking the excavated material and then evaluating whether the broken samples can be further broken into smaller size samples does this.

Unit Ten   **Personal Protective Equipment**

Answers

1. The minimum level of personal protective equipment required for trench rescue operations is all of the following except:

   c. Skullcap

Discussion: All of the other equipment is required as minimal protection. The skullcap, while providing excellent cooling effects and currently very trendy, is not required.

2. Fire fighting gear should never be worn on a trench collapse.

   d. False

Discussion: Environmental or other safety considerations may make the set of typical fire fighting gear the most appropriate. Keep in mind that it is not the preferred gear for trench work.

3. The minimal level of personal protective equipment worn by rescuers at a trench incident is:

   b. Determined by the specific hazards presented.

Discussion: There are any number of different combinations of personal protective equipment that can be worn on a trench incident. Rescuers should be in the clothing and equipment that will protect them from the environment and other incident conditions while also being the most comfortable and easy to wear while working.

4. Eye protection should:

   e. All of the above

Discussion: Eye protection is very important on the trench incident because nails and other protruding objects can cause serious injury. While wrap-around goggles afford more protection, they frequently fog up and therefore hinder the rescuer's vision.

Unit Eleven   Equipment & Tools for Trench Rescue Operations

Answers

1. Ropes that are used to set and adjust panels are best attached to the top of the panels.

    b.   False

Discussion: Adjustment and lowering ropes are attached to the panel by pushing one end of a rope through a hole cut in the panel. By tying a knot in the end of the rope on the strongback side, or looping it around the strongback at the bottom, the panel can be adjusted from the top of the trench. In all cases, the attachment of ropes is at the bottom of the panel.

2. If 4 x 8 ground pads will not fit on the spoil pile side of the trench you should:

    b.   Place 2 x 12s on that side

Discussion: Ground pads should always be used when working around the trench opening. If the volume of soil around the trench lip is too great to place a 4 x 8 piece of plywood, then consider moving enough of the dirt to put in place a 2 x 12.

3. The use of ground pads is primarily for the distribution of vertical force caused by people standing around the trench lip. The main disadvantage to using ground pads is:

    b.   They can cover cracks and separations in the soil

Discussion: Trench rescue events are very dynamic and conditions change rapidly. Ground pads can cover a significant portion of the trench rescue area making observation of the ground under the trench difficult.

4. The shores and installed upright in a lumber trench are the only part of the protective system considered by OSHA for compliance.

    a.   True

Discussion: While the manufactured panels are very strong, they are not considered a part of the protective system when using the OSHA shoring tables. If you are using a shoring system other than that specified by OSHA and it has been certified and designed by a registered engineer, it is acceptable to use the strength of the panels in the system.

5.   A duplex nail is designed to be driven entirely into the wood.

   b.   False

Discussion: The duplex nail is designed to be driven in only to the first shoulder. This will leave a remaining shoulder exposed to pull the nail out.

6.   When using a ladder as a makeshift wale you should make certain:

   d.   All of the above

Discussion: Using a ladder is never the preferred method when constructing a wale. It should be considered only as a last resort; however, in all cases, if you elect to use a ladder, make certain it is fire service grade.

7.   When digging in the trench the preferred tool is the:

   b.   Entrenching tool

Discussion: Digging in the bottom of a trench usually means you are in a very confined area. The smaller and more efficient tool to dig under these conditions is the entrenching tool.

8.   Ladders used in trench rescue can be used as:

   e.   All of the above

Discussion: You need to have multiple ladders on the scene of a trench rescue.

9.   The centrifugal pump can move more water than a diaphragm pump and therefore is the preferred type dewatering device.

   a.   False

Discussion: Mud pumps will only flow low volume, but is more suited for the type of debris-laden water found in a trench.

Unit Twelve   **Air Bags for Trench Rescue**

Answers

1. When lifting in the trench environment you must consider which of the following:

    d. 1,2,3,4,5

Discussion: The weight, height, stability, and surface area of both the object being lifted and the lifting device capacity all need to be considered before effecting a lift. In addition, the surface area under the lifting device needs to be substantial enough to support the load.

2. High-pressure air bags offer the following advantages:

    e. Both a and c are correct

Discussion: High-pressure air bags have greater capacities to lift than low- pressure air bags; however. they cannot lift them as high. Because of the high pressure, they are constructed with puncture resistant material.

3. Low-pressure airbags are usually field repairable.

    a. True

Discussion: Because of the low pressure used to inflate the low-pressure air bag, a leak can be stopped very easily. In some cases where there is a hole and not a tear, the rescuer can place his finger over the hole until the bag can be repaired or replaced.

4. The ability of an air bag to effect a lift is a function of:

    b. All of the above

Discussion: The surface area of the object being lifted, the pressure being supplied to the bag, and the surface area of the bag available to make the lift are just a few of the important considerations when making a lift using air bags.

5.   A four-point (4" x 4") box crib will theoretically support

   d.   24,000 pounds

Discussion: Box crib systems are rated by the number of contact points avail-able to transfer energy. The rule of thumb for 4 x 4 box cribs is 6,000 pounds per contact point. Since there are four points of contact, the total capacity of the four point box crib is 24,000 pounds.

Unit Thirteen   Trench Rescue Assessment

Answers

1.   Clues as to the location of a buried victim include:

   e.   All of the above

Discussion: Clues to locating a buried victim are tied to what the victim was doing at the time of the collapse. Finding the site supervisor is vital in this effort.

2.   The determination of facts and conditions that led to the collapse of a trench is called a:

   a.   Size-up

Discussion: The determination of what has occurred is done by size-up of the situation. A proper size-up will help you determine what equipment and man-power is required to effect the rescue successfully.

3.   Which of the following are important considerations in the size-up of a trench rescue:

   e.   All of the above are correct

Discussion: What has occurred, the work being done, access problems, and equipment and manpower are all important; however, the most important con-sideration is determining whether you are involved in a rescue or recovery.

4. Which of the following are important considerations after arrival on the scene of a trench rescue:

    e. All of the above are correct

Discussion: Finding someone who is in charge, and making sure there is no language barrier, may lead you to the scene's Competent Person. This person should have intimate knowledge of the work being done and the type of protective system in place. Information gathered after arriving on the scene will help you decide if the collapse is within the capabilities of your equipment and team and, in addition, will help determine the victim's survivability profile.

Unit Fourteen   **Hazard Control**

Answers

1. Which of the following is not one of the hazard control categories:

    b. Hydraulics

Discussion: Mechanical, electrical, water, chemical, and man-made are the hazard control categories. Depending on the type of hazard, hydraulics could either be a chemical or mechanical hazard.

2. The control of scene hazards takes place in two phases, the general area and the logistics area.

    b. False

Discussion: The control of scene hazards needs to be eliminated in the general and rescue area. Which hazards you deal with first are determined by the severity of the hazard and the immediacy at which it needs to be controlled to provide a safe rescue scene.

3. When considering shutting down heavy equipment at the scene you should consider:

    e. All of the above are correct

Discussion: There is no standard answer for if a running piece of machinery should be shut down and left in place or removed from the rescue area. In most cases if the machine won't interfere with the protective system or rescue effort it is usually safer to leave well enough alone. In all cases, if you are shutting down a piece of hydraulic equipment and its boom is still under pressure, you should support it with cribbing or chain it so it is in a zero mechanical state.

4. The minimum safe area that should be established around a trench incident is:

    d.   300 feet

Discussion: There are many factors to consider when establishing a safe area around a trench collapse. A general rule of thumb is 300 feet; however, this may have to be smaller or larger depending on your specific scene factors.

5. The location of all underground utilities is important before beginning the rescue effort. For this reason it is a good idea to:

    d.   All of the above

Discussion: Most areas have utility marking services that come out and mark underground utilities prior to any construction or digging project. Make sure your marking service number is included on your emergency contact list.

Unit Fifteen   Atmospheric Monitoring for Trench Rescue

Answers

1. Trenches that are four feet or more in depth must be atmospherically monitored when it can reasonably be expected that an atmospheric problem may exist. This monitoring should be for which of the following:

    b.   Oxygen, flammability, and toxicity

Discussion: Oxygen, flammability, and toxicity should always be monitored before any action on a trench rescue is started. Remember that you always monitor oxygen first because low or high O2 readings could give you an inaccurate flammability reading.

2. When a preset alarm activates in your monitor it indicates:

    b. Some action on your part is necessary

Discussion: If the alarm occurs after the initial set of readings, it would indicate a change in the situation and some action on your part is necessary. If the alarm is on your first set of readings you could very well be in a dangerous situation.

3. When you are using ventilation for hazard control you should:

    c. All of the above

Discussion: There are many factors that need to be considered when choosing to use ventilation for hazard control. Have a monitoring guru on scene to assist you in taking additional readings for determining the success of your ventilation.

4. Your response to a trench incident should include the Haz-Mat team.

    a. True

Discussion: Atmospheric monitoring is a technical expertise that is best left to the folks that can do it best. Have a Haz-Mat response as a part of your initial response to a trench incident.

Unit Sixteen   **Gaining Access**

Answers

1. Which of the following methods concerning digging operations is unacceptable:

    b. Using a backhoe to pull away material that is remote from the victim

Discussion: Hydraulics has no sense of touch or feel. Resist the temptation to do it the easy way unless "parts are ok."

2. When gaining access to a victim of a trench collapse the primary concern is to:

    b. Uncover the head and chest first

Discussion: While this may sound like it doesn't need to be said, in some cases the victim's head may be very close to a protruding foot or hand. The bottom line is to look for the head and chest first so an airway can be established.

3. Accidents that involve a cave-in fall into two different categories:

   b. Those in which the victim is buried or partially buried

Discussion: While one could make a case for the correct answer is being alive or dead, the real issue for us is whether the victim is buried completely or just partially. Generally speaking, the partially buried victim represents a true emergency, while the completely buried victim's survivability profile is not very good and therefore no emergency would exist.

Unit Seventeen   Protective Systems in Trench Operations

Answers

1. Traditional sheeting and shoring operations used by the rescue service are typically only good in trenches up to:

   b. 15 feet

Discussion: When trenches get deeper than 15 feet, the compressive forces are too great for the type of equipment used by most rescue teams. In these cases, commercial trench boxes or sloping would be the preferred technique.

2. The first set of panels placed at the scene of a collapse should if at all possible be placed:

   b. On either side of the victim

Discussion: Because of the secondary collapse potential, the first set of panels should always be placed on either side of the victim. In many cases, even if the shores are not in place, the panels will form a protective zone over the victim as the soil collapses.

3. There has been a massive collapse and you have no walls left that you feel are adequate to support a panel and shore protective system. You may consider:

   d. Both A and B are correct

Discussion: In a situation like a massive collapse, commercial techniques are almost always the best method of handling the situation.

4. When sheeting and shoring a trench we are making an attempt to stabilize the unstable and keep the stable from becoming unstable. We accomplish this by:

   d. All of the above

Discussion: Sheeting and shoring transfers energy from one side of the trench to the other through the strongbacks. This allows the pressure to be distributed and takes away the soil's active potential.

5. Inside waler systems are used for a number of reasons. Considerations in the use of inside wales would be to:

   d. All of the above

Discussion: Inside wales can be used to create an open space if the shores in the middle of the protective system would hinder victim removal. This type of system requires fewer shores and, depending on the number of strongbacks used, can create more contact points for energy transfer.

6. If the stability of your protective system is in question, consider which of the following options:

   e. All of the above

Discussion: There are many techniques for increasing the strength of your protective system. In general inside wales with additional strongbacks will create additional contact points for energy transfer. In any situation you should not be shy about calling for an engineer who can assist in designing an adequate system for the situation.

7. The procedure for installing pneumatic shores is:

    c. Middle, bottom, top

Discussion: Because pneumatic shores can be installed from outside of the trench, they are placed in the weakest part of the trench wall first. In general, if you can get a middle shore shot in a 10-foot trench, it will be sufficient to keep a secondary collapse from occurring.

8. The procedure for installing the initial set of timber shores is:

    a. Top, middle, bottom

Discussion: Timber shores must be manually installed and therefore the top shore is installed first. The rescuer should never be in a trench at any time with their waist lower than the last installed shore.

9. When digging operations involve the removal of more than _____ feet of soil below the strongback or panel, you are required to provide supplemental sheeting and shoring protection.

    c. 2

Discussion: The removal of more than 2 feet of soil below the strongback requires that supplemental shoring be installed.

10. The opposite side panel set is used to:

    d. Both b and c are correct

Discussion: The opposite side-panel set technique allows the panel team to set the panels from the opposite side of the trench. This is an excellent technique for situations when access to one side of the trench is not feasible or when there is a victim in the trench and a very precise set is necessary.

11. During shoring operations the shooting of the shore is accomplished by a command given:

    c. By the shore installer

Discussion: Since the shore installer is the one with his fingers in and around the shore, he is the only one who should be calling for the shore to be shot.

12. When placing panels at any trench collapse, you should always strive to create a safe working area by providing a minimum of ____ feet of protected area:

    b.   12 feet

Discussion: Taking for granted that your first set of panels is directly over the victim, the rescue team should strive to create a safe area on either side of the victim. Since the panels are four-feet wide, the total width of three panels would be 12 feet.

13. The maximum distance between horizontal shores should not exceed:

    b.   4 feet

Discussion: Regardless of the distance requirements for vertical shore spacing, unless wales are used, the maximum distance between shores horizontally should be four feet. This would allow each cone of pressure to overlap within the trench walls.

14. Which of the following is not true about isolation tunnels:

    c.   55 gallon-welded steel drum can be used horizontally

Discussion: Drums or barrels can be used as isolation devices by cutting the ends off and using it in a vertical fashion. They should never be used horizontally as pressures may be too great and cause the barrel to collapse. If you are using concrete or steel pipes they can be installed horizontally.

15. Sloping and benching is a protective system method approved by OSHA to provide safety for workers in a trench.

    a.   True

Discussion: Sloping and benching are approved techniques for trench protection. The soil class determines the height to length ratio that would create a safe area.

16. There is a variety of shoring systems available on the market today. When using a pneumatic shore that has a T-handle device or Acme thread, the user can exert additional levels of pressure on the wall by manually turning and securing this device. These pressures are:

    c.   Up to and in some instances more than 450 psi

Discussion: The inclined plane design used to tighten some shores can create very high forces and in some cases far greater pressures than they can be shot manually. What this means is that if shores need to be retightened, it is not necessary to reattach the air line. They shore installer can just perform this function manually.

17. The first shoring team member to enter the trench for placement of shores should be in:

    c.   Class 3 harness with tag line

Discussion: As the first set of panels and shores are assembled to create a safe area, all rescuers working around the trench opening and subject to falling in the trench should be in a Class 3 harness so they can be pulled from the trench, using a rescue rope if necessary.

18. Commercial techniques should be considered if:

    f.   All of the above are correct

Discussion: Trenches over 15 feet deep, situations where there is a massive cave-in, running debris, or environmental factors that are beyond your capability to correct may well require commercial techniques. Never be afraid to call in the people who do construction for a living. They may not understand the rescue system but they are usually very proficient in commercial techniques.

Unit Eighteen   Victim Packaging & Termination Procedures

Answers

1. Considerations for victim packaging should begin when the victim is uncov-
    ered.

    b.   False

Discussion: The extrication effort and plan to remove the victim should be in
place early in the incident. The reason for this is that the design of the protective
system must take into account victim access and removal.

2. During the removal of a victim from a trench collapse your greatest concern(s)
    should be:

    d.   All of the above

Discussion: The victim packaging device you choose should be one that will
protect the victim from further injury and also fit the existing exit route from the
trench. The key is to get the victim to definitive medical care as quickly as practical.

3. If a lifting operation were necessary on a trench accident it would most
    appropriately be handled by:

    d.   Rigging Team

Discussion: If there are enough people on the scene, you may want to establish
a rigging team to build an elevated point of attachment and rig a mechanical
advantage system to help with victim removal.

4. Termination can be the most hazardous time of the incident.

    a.   True

Discussion: When the event excitement and adrenaline have faded, rescue
personnel may have a tendency to relax. Don't let them get hurt during this
phase of the operation because they are tired and stop concentrating.

Unit Nineteen   Techniques for Trench Protection

Answers

1. The minimum number of panels required for a T-trench protective system is:

    a. 7

Discussion: The seven panels are used in conjunction with inside wales to span a center panel on top of the T.

2. Your primary consideration during panel placement and shoring on a T-Trench is:

    c. Capture and control the corners

Discussion: As with the L-Trench, the capture of the exposed corners is of primary concern.

3. In the L-Trench "thrust blocks" are required on all panels.

    b. False

Discussion: Thrust blocks are only required on the panels that will have shores shot to the corner blocks.

4. When constructing the L-Trench, there are several force/counter force issues that need to be pre-planned for and engineered into the system. These include, but are not limited to:

    b. 1, 2, and 3

Discussion: Since some shores will be shot to an angled thrust block, care must be taken not to provide so much pressure that the shore pushes the panel over. The angles of the thrust block should be shot after the straight shores are set and to somewhat less pressure.

5. What is the primary consideration when creating a system to render an L-Trench safe?

    b. Control the inside exposed L walls

Discussion: The most hazardous portion of the L-trench is the inside corner of the L. Our first efforts at stabilization should be to capture it with panels and shores.

6. After the installation of pneumatic shores they should be toe nailed to:

    b. Keep the shore from falling out if it becomes loose

Discussion: Since all pneumatic shores when finished are pinned or locked manually, they can become loose or dislodged as other shores are set. Toe nailing the ends of the shores will keep them in place until they can be retightened manually.

7. Ventilation is required on all trench incidents.

    b. False

Discussion: Ventilation is only required as the situation dictates. If there are environmental conditions that prohibit ventilation and no atmospheric problems exist, then it is not necessary to ventilate the trench.

8. In the single wall slough trench, the technique to use would be:

    d. Outside wales

Discussion: Because the slough leaves a gap in the trench wall it is necessary to span the gap using outside wales. Just don't forget to backfill the void area so that proper transfer of forces can take place.

9. In order to create an open space in a trench, the technique would be:

    b. Use inside wales

Discussion: Inside wales are used to create an open space because the middle set of panels is spanned by the wales and no shores are required to be shot to those panels. The end result is a very open working area in which to dig or package a victim.

10. During deep trench operations (15-foot trench) it is acceptable to turn the top panels horizontal and shoot them first.

    a. True

Discussion: The length of the strongbacks prohibits the panels from both going in vertically. The top set, which will have to be set first, has to be turned so that the strongback is horizontal.